RED ROCKS COMMUNITY COLLEGE

D0046866

077257

QB 521 .S675 2004

Spence, Pam.

Sun observer's guide

DATE DUE

GAYLORD 234   PRINTED IN U.S.A.

FIREFLY

SUN

OBSERVER'S GUIDE

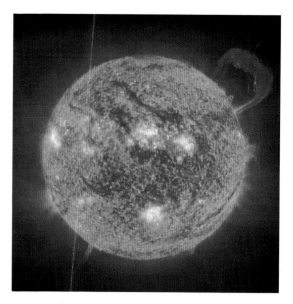

PAM SPENCE

RED ROCKS
COMMUNITY COLLEGE LIBRARY

FIREFLY BOOKS

*61331899
077257
QB
521
·S675
2004

A FIREFLY BOOK

Published by Firefly Books Ltd. 2004

Copyright © 2004 Pam Spence

All rights reserved. No part of this publication may be reproduced, stored in a retrieval system, or transmitted in any form or by any means, electronic, mechanical, photocopying, recording or otherwise, without the prior written permission of the Publisher.

First printing

Publisher Cataloging-in-Publication Data (U.S.)

Spence, Pam.
   Sun observer's guide / Pam Spence. —1st ed.
[160] p. : col. ill., photos. ;   cm.
Includes bibliographical references and index.
Summary: A guide to observing the sun for amateur astronomers.
ISBN 1-55297-941-5 (pbk.)
1. Solar activity--Observations. 2. Sun--Amateurs' manuals.
I. Title.
522 22   QB524.G37 2004

National Library of Canada Cataloguing in Publication

Spence, Pam
        Sun observer's guide / Pam Spence.
Includes bibliographical references and index.
ISBN 1-55297-941-5
        1. Sun--Observers' manuals. I. Title.
QB521.S64 2004   523.7   C2004-900412-3

Published in the United States in 2004 by
Firefly Books (U.S.) Inc.
P.O. Box 1338, Ellicott Station
Buffalo, New York 14205

Published in Canada in 2004 by
Firefly Books Ltd.
66 Leek Crescent
Richmond Hill, Ontario L4B 1H1

Published in Great Britain in 2004 by Philip's,
a division of Octopus Publishing Group Ltd,
2–4 Heron Quays, London E14 4JP

Printed in China

Front Cover and Title Page:
The Sun showing a
prominence at upper right.
(Courtesy of SOHO/EIT
consortium. SOHO is a
project of international
cooperation between ESA
and NASA.)

Back Cover: (top)
Diagram of solar rotation.
(bottom) Projecting the
Sun's image through a
star diagonal.

CONTENTS

RED ROCKS
COMMUNITY COLLEGE LIBRARY

WARNING
Never look directly at the Sun unless using appropriate and undamaged filters. See the guide to safety measures on page 36.

WHY OBSERVE THE SUN?

The Sun is a star – a huge, massive body generating energy from nuclear reactions in its core. Its importance to life on Earth cannot be underestimated: without the Sun, we would not exist. However, the Sun can also be harmful, and it needs to be monitored closely if we are to have the opportunity of forecasting any changes which might affect the Earth. Solar astronomy is one of the few remaining areas of astronomy in which amateurs can make observations which are of use to professionals, without the need for expensive equipment. All that is required is a small refracting telescope and a cardboard box!

This book is the ideal introduction to observing the Sun. It gives clear, step-by-step instructions on how to project the Sun's image safely using a small telescope or just binoculars. It explains the significance of what can be observed, and how observations can be made for personal enjoyment or for submitting to professional organizations, and describes how to photograph the Sun. There are chapters on the Sun's structure, the Sun–Earth interaction, solar eclipses, and professional solar astronomy conducted via spacecraft. An extensive glossary gives definitions of scientific terms used in the book.

The Sun dominates the Solar System, and also dominates life on Earth. Without the Sun's heat, light and energy, life on our planet would be impossible. All energy comes from the Sun – even the fossil fuels we burn originally got their energy from the Sun. Luckily for us, the Sun will continue to give out heat and light for billions of years to come, but our environment exists on a knife-edge, and hence so does our survival. The Earth's atmosphere protects us from harmful solar radiation, but we are already having a detrimental effect on our protective blanket; we create chemicals and gases which break down protective layers in the atmosphere, and we are cutting down the vegetation that removes harmful carbon dioxide from the air.

The Sun's radiation output varies on a daily basis and also over longer periods of time. By observing the Sun we can help to predict the effect of solar activity on the Earth. The Sun's activity impinges on the Earth in many ways, from interfering with our communication systems to initiating major climate changes. There is still a great deal that is not understood about how the Sun affects the Earth, so by daily observing we can hope to increase our knowledge. Solar observing is one branch of astronomy where amateurs can make a significant contribution to the science. It is also an activity for which it is not necessary to possess the latest, largest and most expensive equipment. Indeed, smaller telescopes are better than larger ones for observing the Sun.

Observing the Sun is not difficult, but it does pose challenges. One aspect which has to be addressed is safety. The Sun is a very powerful source of light, heat and radiation. Even glancing at the Sun with the naked eye can cause permanent and serious damage to the eyes. If the Sun is observed directly without proper protection, blindness can result. The importance of observing safely cannot be overemphasized. Once the danger of the Sun is acknowledged and it is treated with the respect it deserves, solar observing can be made 100 per cent safe. By following simple and sensible precautions the danger is entirely removed. *Never* look directly at the Sun, with or without magnification, unless using adequate, safe and well maintained filters. *Never* leave any instrument pointed at the Sun unattended. *Always* supervise children and the general public when they are near solar observing equipment. Further safety measures are given at the beginning of Chapter 3.

For obvious reasons, the Sun has always been of huge significance to humans. In ancient cultures the Sun was often the dominant god. When a solar eclipse cast its shadow, people were terrified and would do anything to appease the gods to bring the Sun back. The Sun was widely observed in ancient times. Its position in the sky dictated the yearly calendar and, astrologers believed, affected human affairs. There are records of sunspots being seen in ancient times. In the nineteenth century astronomers began to make daily records of the number of sunspots on the solar disk, but it was not until the twentieth century that their nature and significance were realized. A variety of solar phenomena are closely associated with sunspots, including solar "storms" – huge outpourings of energy which, when it reaches the Earth, can knock out power grids over vast areas, fry satellites, and interfere with telecommunication systems.

Today, sophisticated instruments on board spacecraft monitor the Sun at all wavelengths of radiation, helping to predict when solar storms might strike, but solar observing is not just for the professionals or sophisticated spacecraft. Some of the closest collaboration between amateur and professional astronomers happens in solar astronomy. This is a field which is open to everyone. By following simple instructions, the amateur astronomer equipped with just binoculars or a small telescope can make observations which, submitted to an observing organization, can help the professionals uncover the secrets of the Sun. Observing the Sun is great fun and intriguing. There are few astronomical objects that change continually over the course of a few hours, few for which true and important science can be done with a minimum of equipment, and even fewer that can be observed outside the hours of darkness.

— I · THE STRUCTURE OF THE SUN —

The Sun is a star, shining as a result of energy produced within it. Many objects, including planets such as the Earth, can be seen only by the light they reflect from stars. For the Earth, of course, it is the Sun's reflected light that allows it to be seen. The Sun is the ultimate source of all light and energy on the Earth. The radiation from the Sun takes approximately 8.3 minutes to travel the distance of about 150 million km to the Earth. We thus observe the Sun as it was 8.3 minutes ago.

The Sun is composed mostly of hydrogen (71% by mass) and helium (27%), the other 2% consisting of heavier elements, mainly carbon, nitrogen and oxygen. It produces its energy deep within itself, in the core, where the solar material is at its most dense and at its highest

The electromagnetic spectrum

Light is a form of radiation. In 1865, the Scottish physicist James Clerk Maxwell showed that electric and magnetic forces are two aspects of the same phenomenon: electromagnetism.

Radiation consists of an electric and a magnetic field oscillating perpendicular to each other. This electromagnetic (e.m.) radiation can be thought of as tiny packets of energy called photons. They travel as waves but interact like particles. The more energetic the radiation, the higher the energy of the photons and the shorter the wavelength of the wave. Through empty space (a

vacuum) e.m. radiation travels at a speed of just under 300,000 km/s.

We view the world in visible light. Photons with energies corresponding to the wavelengths of light are emitted by the Sun; when they reach the Earth they bounce off objects into our eyes, and our brains construct images of those objects. We are familiar with many other kinds of e.m. radiation: X-rays, infrared, microwaves and radio waves. These are all types of e.m. radiation, differing only in their energy and wavelength. The whole range of electromagnetic radiation is known as the electromagnetic spectrum, illustrated below.

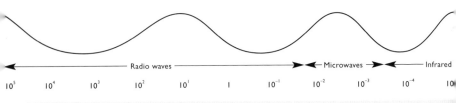

| | | Radio waves | | | | | Microwaves | | Infrared |

10^5 10^4 10^3 10^2 10^1 1 10^{-1} 10^{-2} 10^{-3} 10^{-4} 10

temperature. The temperature in the core reaches 15 million K (degrees kelvin) and the density reaches 1.6×10^5 kg/m^3, while the pressure is about 200 billion times that of the Earth's atmosphere at sea level. At these temperatures and pressures, the hydrogen is transformed into helium by nuclear fusion – often termed hydrogen burning. In each fusion reaction, four hydrogen nuclei are changed into one helium nucleus and a tiny amount of energy is released. The Sun, which has a mass of 2×10^{30} kg (330,000 times the mass of the Earth), converts 6 hundred thousand million kg (6×10^{11} kg) of hydrogen into helium every second, which explains why its energy output is prodigious. The total power output of the Sun (the amount of energy radiated out into space every second) is 3.86×10^{26} watts. It has been burning hydrogen in its core for 4.6 billion years and will continue to do so for another 5 billion years.

The Sun emits radiation across the whole of the electromagnetic spectrum, from the very high-energy, short-wavelength gamma rays to the low-energy, long-wavelength radio waves. However, most of the radiation – apart from the small part we call visible light, and some radio wavelengths – fails to reach the Earth's surface. The short-wavelength radiation is absorbed by the Earth's atmosphere, while the longer-wavelength radiation bounces back into space. It is thus no accident that our eyes have evolved to respond to visible light.

Visible light is also termed white light. If light from the Sun passes through a prism or through raindrops, it produces a rainbow of colors. This familiar spectrum is produced because everyday white light is composed of radiation of different wavelengths. When white light passes through a prism or raindrops, the different wavelengths are refracted (bent) at different angles and spread out to form a rainbow.

Observing in different parts of the electromagnetic spectrum reveals different information: for example, taking an X-ray of a limb shows the bones, while an ordinary photograph shows the skin.

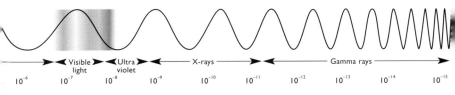

| Visible light | Ultra violet | X-rays | Gamma rays |

10^{-6} 10^{-7} 10^{-8} 10^{-9} 10^{-10} 10^{-11} 10^{-12} 10^{-13} 10^{-14} 10^{-15}

Energy transportation

The energy produced in the Sun's core is in the form of gamma rays (see the feature on pages 6–7, "The electromagnetic spectrum"). This energy diffuses very slowly outward, the gamma-ray photons losing energy as they travel. Since the density at the core is so high, a gamma-ray photon does not travel very far before it is absorbed by another particle; this particle will then emit another gamma ray, perhaps inward toward the core, perhaps outward. Slowly the energy is transferred outward; it can take a million years for photons to diffuse out from the core to the top of the photosphere. Each time they interact with particles, they lose energy, and as they move away from the core, the pressure and temperature around them fall. The layer in the Sun where this radiative diffusion dominates the outward flow of heat is known as the radiative zone.

In the outer layers of the Sun, the main mechanism of energy transfer is convection. This is how heat travels from the bottom of a pan of porridge to the top. The heat from the stove raises the temperature of

▼ The Sun's energy is created in the core. It travels outward by radiative diffusion in the radiative zone, then by convection in the convective zone.

The layer that we see with our eyes is the photosphere. Beyond the photosphere lie the chromosphere and corona.

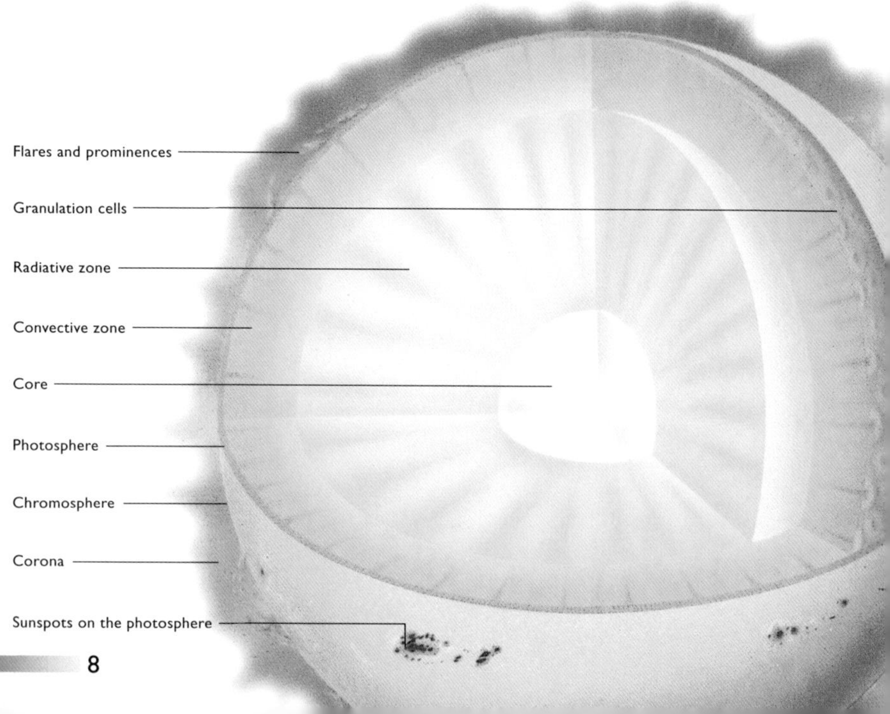

Flares and prominences

Granulation cells

Radiative zone

Convective zone

Core

Photosphere

Chromosphere

Corona

Sunspots on the photosphere

▶ *The photosphere is at the top of the Sun's convective zone, so just like the top of a heated pan of porridge, the surface shows the effect of convection. This granulation can be seen in ordinary light through quite small telescopes on very clear days.*

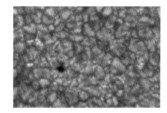

the porridge at the bottom of the pan, which then rises and expands, cooling as it moves away from the source of heat. Eventually it reaches the top of the pan and sinks back down to be heated again. Gradually the porridge is heated all the way through, and in the process heat is transferred from the stove to the top of the pan. In much the same way, solar material is heated and rises to the top of the convective zone, where it cools and sinks again. The layer of the Sun where convection dominates the outward flow of heat is called the convective zone. The tops of the convection cells can be seen in white light, even with a small telescope, as what is known as granulation – a network of small, lozenge-shaped cells.

The solar "surface"

The outer layer of the Sun, at which the temperature falls to around 5500 K, emits energy in the form of visible light. This layer, called the photosphere (meaning "sphere of light"), is sometimes regarded as the "surface" of the Sun. Before the nature of the Sun was discovered, people seriously thought that the Sun's photosphere was solid (William Herschel, the famous 19th-century astronomer, for example, thought that sunspots were land spied through cloud). We now know that this is not a solid surface as the Sun is not a solid body, but it is the part of the Sun we can see with our eyes.

The Sun does look like a smooth, solid body when we look at it in ordinary light, because the photosphere is only about 500 km thick. Compared with the diameter of the Sun, which is 1,390,000 km (over a hundred times the Earth's diameter), this is a very thin layer. It appears smooth because any fluctuations within it are too small for us to discern. But if we observe the Sun in other wavelengths of radiation, for example in X-rays, we see that it is a dynamic body, pouring out material from the photosphere.

The outer regions

Strangely, the material in the Sun above the photosphere gets hotter rather than cooler, as expected. The source of this heating is not fully understood, but it is believed the Sun's magnetic field is involved in

Inside the atom

To understand solar phenomena, we have to look at what happens within the atom because processes in the Sun take place under extreme conditions. At such high pressures and temperatures, the constituent parts of atoms can exist independently, allowing events to happen at a subatomic level that cannot do so naturally on Earth.

Every atom is composed of a nucleus, made up of one or more protons and (except for hydrogen) neutrons, around which electrons orbit. Electrons are much lighter than protons and neutrons, and have a negative electrical charge. Protons and neutrons each have a mass 1840 times the mass of the electron, but while protons have a positive electrical charge of the same amount as the electron, neutrons are electrically neutral – with no charge at all.

Each chemical element is uniquely defined by having different numbers of protons, and the different subatomic composition – in particular, the number of electrons – dictates the chemical behavior of the element. Different isotopes of the same element have the same number of protons, but different numbers of neutrons.

Electron orbits

Electrons do not orbit the atomic nucleus randomly: only certain orbits are allowed, each corresponding to a different energy. The numbers of electrons allowed in each orbit are also restricted. If an orbit has its quota of electrons, then any more electrons have to occupy a higher orbit.

The different orbits have discrete energies, so if an electron moves from one orbit to another, it has to lose or gain the exact amount of energy corresponding to the difference in energy levels. Electron energies are said to be quantized. The energy gained or lost by an electron moving between orbits is absorbed or emitted as a photon (see the feature "The electromagnetic spectrum" on pages 6–7).

Hydrogen and helium

The Sun is primarily composed of hydrogen and helium. Hydrogen (H) is the lightest chemical element and the most abundant element in the Universe. Its commonest isotope consists of one proton orbited by one electron, with no neutrons in the nucleus. Two other isotopes exist: deuterium, with a nucleus of one proton and one neutron, and tritium, with a nucleus consisting of one proton and two neutrons. Deuterium is stable, but tritium is unstable and radioactively decays (into an isotope of helium).

Helium (He) is the second-lightest chemical element. It was actually discovered in the Sun in 1868, independently by the French astronomer Jules Janssen and the English scientist Norman Lockyer, 30 years before being found on Earth. It was named for Helios, the Greek Sun god. In its most abundant isotope, helium consists of a nucleus composed of two protons and two neutrons, about which two electrons orbit. This very stable nucleus is called an alpha particle. Three other isotopes exist; two are unstable, but the third is stable with a nucleus consisting of two protons and one neutron.

Different forms

As well as having different isotopes, elements can exist in different forms: atomic, molecular and ionized. Molecules are groups of atoms that can exist in a stable form together. Molecular hydrogen (H_2) consists of two atoms of hydrogen. The atomic form is electrically neutral. Atomic, or neutral hydrogen is sometimes denoted by HI.

When an element is ionized, it loses or gains electrons, which affects the overall charge of the atom. If it gains an electron, the atom becomes negatively charged. If it loses electrons, it becomes positively charged. The hydrogen in the Sun, because of the high temperatures, is generally ionized hydrogen, HII. The electrons have too much energy to be captured by a proton to form a hydrogen atom. Matter where all or most atoms are ionized is called a plasma.

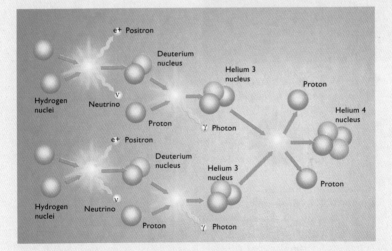

▲ Within the Sun hydrogen is being changed into helium by the process of nuclear fusion. During this process four hydrogen nuclei merge into one helium nucleus and a tiny amount of energy is emitted.

◄ *The corona is the outer atmosphere of the Sun. Its pearly white light is a beautiful sight during totality of solar eclipse. At other times the photosphere drains out the fainter light of the corona.*

◄ *The dark area on the disk of the Sun is a coronal hole. Solar material from coronal holes streams out into the solar wind, which in turn streams out into interplanetary space carrying the solar particles.*

some way. The temperature rises from the bottom to the top of the chromosphere – the region of the Sun immediately above the photosphere. The chromosphere ("sphere of color") can be seen as a narrow ring of brilliant red at the time of a total solar eclipse. It appears red because the hydrogen atoms in it are excited by the high temperatures and they emit radiation in the red part of the visible spectrum (see the feature, "Inside the atom"). The temperature of the chromosphere rises from around 6000 K at its base, which is the average temperature of the top of the photosphere, to 50,000 K at the top.

The Sun's outer atmosphere is the corona, which can be seen during the totality of a solar eclipse. Its ghostly white, almost ephemeral light is a beautiful sight during an eclipse, but at other times it is drowned out by the bright light from the photosphere. The corona extends many millions of kilometers out from the photosphere and can reach temperatures as high as a million degrees K. Its shape varies according to how active the Sun is. At peak activity the strong magnetic fields constrain the corona largely to the equatorial regions, whereas at low solar activity the corona can be seen all around the solar disk except at the extreme poles.

Observing the Sun in X-rays reveals the corona in detail, showing vast "holes" that can appear in the outer atmosphere. It is thought that these coronal holes are lower-density regions where magnetic field lines stream out from the Sun into interplanetary space. There appear to be two permanent coronal holes, with one centered at each pole.

The solar wind

At the very high temperatures within the Sun, the solar material is ionized. This means that many electrons have too much energy to be captured by atomic nuclei, so the material exists as what is termed a plasma – a "sea" of negatively charged electrons and positively charged nuclei or ions (see the feature on pages 10–11, "Inside the atom"). The movement of these charged particles creates a magnetic field, and this magnetic field plays a very dominant role in the many aspects of solar activity.

The Sun is continually emitting particles into space. This outward flow of charged particles (mainly electrons and protons) is termed the solar wind. The solar wind streams out through the Solar System,

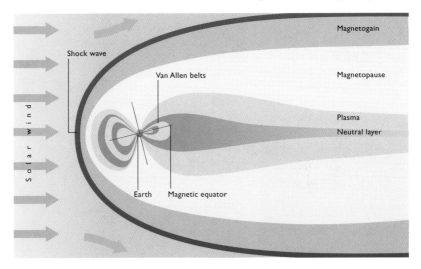

▲ The Earth's magnetic field is flattened on the sunward side by the solar wind; on the other side of the Earth it is drawn out into a tail. The electrically charged particles in the solar wind cannot easily cross the magnetic field lines of the Earth and a shock wave is generated.

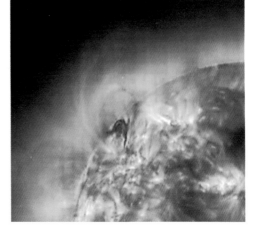

▶ *The SOHO spacecraft captures the eruption of a coronal mass ejection (CME). A billion tons of solar material is blasted into space.*

interacting with all the planets, including the Earth. Near the solar equator the solar wind flows out at around 400 km/s, but the speed of the solar wind emanating from the polar coronal holes is nearer to 750 km/s.

Charged particles such as electrons and protons cannot easily cross magnetic field lines, so the Sun's magnetic field constrains the flow of the solar wind. Magnetic field lines stretch outward from coronal holes, and solar material is able to escape more easily along these lines than from other parts of the solar disk. This creates "gusts" in the solar wind as the material escaping from coronal holes on the rotating Sun sweeps out across the Earth's orbital path. Material emanating from coronal holes crosses the Earth with a period of about 26 or 27 days.

The Sun also emits huge bubbles of plasma during solar flares. Solar flares are enormous releases of solar energy. They erupt near sunspots and are caused by the release of magnetic energy stored in the convoluted magnetic field lines within the sunspot groups when these lines occasionally break and reconnect. The energy released in a solar flare can be equivalent to that of several billion megaton

▼ *The white patches on this image of the Sun are flares. A tremendous amount of energy is released in a solar flare.*

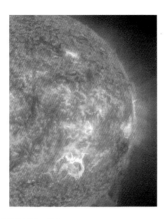

nuclear bombs (up to 10^{25} joules). They eject streams of charged solar particles into the solar wind, creating gusts which, if ejected earthward, can cause problems on Earth (see Chapter 8). The bubbles of plasma released during a solar flare expand out through the corona into interplanetary space. They are called coronal mass ejections (CMEs).

Different suns

The appearance of the Sun depends on what kind of radiation it is observed in. For many years, both amateur and professional astronomers could only observe in white light – the type of radiation to which our eyes react (see the feature on pages 6–7, "The electromagnetic spectrum"). Today, spacecraft observe the Sun continually in different wavelengths, monitoring its various features to give us warning of events that can affect the Earth. The Earth's atmosphere filters out many wavelengths of radiation, but modern technology now allows even amateurs to observe in some radiation other than white light. The instrumentation available to amateurs is described in Chapter 2.

The Sun in white light

White light reveals the photosphere. In even a small telescope, on days of good seeing, the tops of the convection cells can be seen. This is the granulation (from the Latin *granulum*, meaning "grain"). Granulation appears as a mottling of the Sun's disk: bright lozenge-shapes with darker, cooler edges. Granules are about 1000 to 2000 km in diameter. Just as the surface of a bubbling pot of porridge is constantly changing, the granulation cells in the photosphere dissolve and reform over a period of about ten minutes. The main features visible in white light, however, are sunspots.

Sunspots

Sunspots appear dark against the brighter photosphere because they are cooler. The photosphere has a temperature of around 5500 K, while sunspots can be 1000 to 2000 K cooler. Larger sunspots have two distinct shadings, with a gray penumbra surrounding a cooler, darker umbra. Filamentary structure visible in the penumbra first suggested to astronomers that sunspots were regions of high magnetic field. While the Sun has an overall magnetic field of around 1 G (gauss), the field in sunspots can reach 2000 G. These high magnetic fields restrict the flow of heat into these areas, keeping them cooler than the surrounding photosphere.

Sunspots range in size from tiny pores about 1000 km across to complex groups covering billions of square kilometers. Some of the larger ones are easily visible to the naked eye (though even glancing at the

▲ *The white and dark spots in the magnetogram show regions of different magnetic polarity in the Sun. The larger spots are regions surrounding sunspots.*

Sun in the sky can damage your eyes, so filters should always be used), and records from ancient cultures show that the Egyptians, Chinese and Babylonians observed them.

When Galileo and his contemporaries first viewed sunspots through telescopes, there was a great debate as to whether they were part of the Sun or some phenomena in the Earth's upper atmosphere. Many were reluctant to admit that the Sun apparently had acne as it went against the doctrine of Aristotle (still prevailing even more than 1600 years after his death) that objects in the heavens, being nearer to God, were pure and unchanging. Eventually, however, astronomers admitted that these spots must be on the Sun, and they used their movement to determine the Sun's rotation rate.

Spots usually occur in pairs or as part of groups because of the associated magnetic field. The strong magnetic field lines arch outward from one spot and back down into the Sun through the other. This gives the pair of spots different magnetic polarities: the spot where the field lines arch outward has north polarity, and the spot where the lines re-enter the photosphere has south polarity. Sunspot groups have a much more complex magnetic field structure. Today the magnetic fields associated with sunspots can be seen in solar magnetograms.

The solar cycle

In 1843, the German astronomer Heinrich Schwabe noticed that the number of sunspots visible on the solar disk varied over a period of about 11 years. This period is known as the solar cycle. When the Sun is at its most active, at solar maximum, more sunspots appear. At solar minimum, days can pass when no sunspots are visible at all. The cycles of activity on the Sun have been observed for nearly 200 years, and scientists have established an easy empirical method for quantifying solar activity based on observing sunspots (see Chapter 5). The historical record of sunspot counts and calculated relative sunspot numbers constitutes the longest scientific database in existence.

At the beginning of a solar cycle, sunspots appear quite far from the solar equator, though they are rarely seen higher than latitudes of 40°

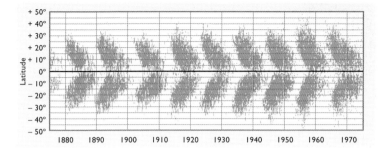

▲ The Maunder butterfly diagram shows the latitudes at which sunspots appear on the Sun's disk. The black line at 0° denotes the solar equator. As a solar cycle progresses, spots appear nearer and nearer to the equator.

north or south. As the cycle continues, however, sunspots appear closer and closer to the equator. Often spots from a new cycle will start to appear at high latitudes while spots from the old cycle are still appearing near the equator. This distribution of spots during a solar cycle can easily be seen in a Maunder butterfly diagram. This graphical representation, named for its shape, was first demonstrated by the English astronomer E. W. Maunder in 1904.

During a solar cycle, all the sunspots leading in the direction of the Sun's rotation in the northern hemisphere will have the same magnetic polarity, with the leading sunspots in the southern hemisphere having the opposite polarity. At the change of a cycle, the entire magnetic polarity of the Sun reverses, though sometimes one hemisphere will reverse its polarity before the other. For the next cycle, the leading sunspots will have the opposite polarity from that which they had before. This "magnetic" solar cycle thus has a period of around 22 years.

The theory of sunspots

There is no full explanation of sunspots and the solar cycle, but the generally agreed scenario is based upon work developed by the American astronomer Horace Babcock in the 1960s. He explained many of the features of the solar cycle in terms of the magnetic field.

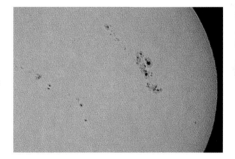

▲ The darker umbra and lighter penumbra of sunspots can be seen in this white light image of the solar disk.

The theory of sunspots

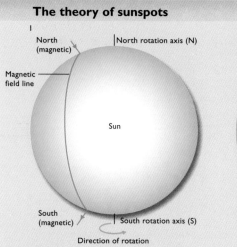

1

North (magnetic)

North rotation axis (N)

Magnetic field line

Sun

South (magnetic)

South rotation axis (S)

Direction of rotation

▲ The Sun has a weak overall magnetic field. It is a plasma, and the magnetic field lines are "frozen" into the solar material.

2

North (magnetic)

N

Sun

South (magnetic)

S

Direction of rotation

▲ As the Sun rotates, the magnetic field lines are dragged around. They are stretched more at the equator because of the differential rotation.

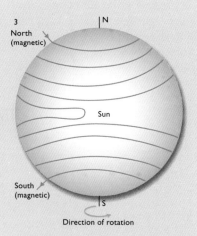

3

North (magnetic)

N

Sun

South (magnetic)

S

Direction of rotation

▲ As the Sun rotates, the magnetic field lines are "wound up" round the disk and stretched out.

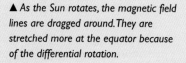
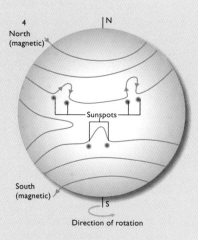

4

North (magnetic)

N

Sunspots

South (magnetic)

S

Direction of rotation

▲ Convection causes the field lines to arch in and out of the photosphere – sunspots are seen at these points.

The Sun's magnetic field of around 1 G is comparable to the Earth's, which is about 0.3 G at the equator and 0.6 G at the poles. The Earth's magnetic field is caused by a molten iron core moving as the Earth rotates, and the field lines created by this "dynamo" effect are similar to those of a bar magnet. The field lines enter and leave the Earth at the poles. The Sun's field, however, originates in the movement of its plasma. Since a plasma is composed of charged particles (negatively charged electrons and positively charged ions), any motion will generate a magnetic field.

The Sun rotates differentially, which means that its rotation speed is faster at the equator than near the poles. The material at the equator rotates in about 27 days, while the material near the poles can take up to 40 days to complete one rotation. As the solar material moves, any magnetic field will be pulled along with it, as charged particles cannot easily cross magnetic field lines. This has the effect of "winding up" the magnetic field.

In addition to the relative horizontal motion resulting from differential rotation, there is vertical movement in the outer layers of the Sun produced by convection. Material from lower levels rises to the photosphere, cools and sinks back down. Again, any magnetic field line will be carried with the movement of the plasma. Thus magnetic field lines near and in the photosphere become distorted and jumbled.

These two motions have the effect of twisting magnetic field lines together, and in these regions the magnetic field increases up to 2000 G. This strong magnetic field restricts the movement of charged particles in these regions, effectively stopping the flow of heat by convection. Hence these regions are cooler than the surrounding photosphere, and they appear dark on the solar disk. The twisted tubes of magnetic field lines arch in and out of the photosphere, and as we look from above, we see sunspots at the place where they break the surface.

This theory explains the existence of sunspots as regions of cooler solar material with high magnetic fields. The theory also explains why the sunspot pairs have different magnetic polarity and why the leading spot has the same magnetic polarity as others in the same solar hemisphere.

The winding up of the magnetic field also explains why sunspots appear at high latitudes at the beginning of a solar cycle, then appear nearer and nearer the Sun's equator as the cycle continues. It takes about 11 years to "wind up" the solar magnetic field, after which there is a build-up of opposite magnetic polarity at the equator and the poles, causing the magnetic polarity of the hemisphere to change – this is the magnetic solar cycle.

There are, however, some aspects of the solar cycle that the theory does not explain, for example the disappearance of sunspots for long periods (see "The mystery of the disappearing sunspots").

Faculae

Also visible in white light are faculae (Latin for "little torches"): bright hot clouds of solar material lying above sunspots in the upper

The mystery of the disappearing sunspots

The relative sunspot number has been used to plot the activity of the Sun daily since 1848, when the Swiss astronomer Rudolf Wolf first suggested using an empirical method to count the number of sunspots visible (see Chapter 5). Wolf and others have subsequently estimated the activity of the Sun before 1848 by looking through historical records. E. W. Maunder concluded that solar activity almost completely died out during the period between 1645 and 1715, a period now called the Maunder minimum.

There was a great debate when Maunder published his findings. Many astronomers considered the past records insufficiently substantial for such a conclusion to be drawn, but independent research on the carbon-14 content of tree rings later confirmed the existence of the Maunder minimum.

Astronomy is rich with seemingly totally irrelevant findings proving or disproving theories. In this case, the carbon-14 is produced when cosmic rays (highly energetic particles, produced by various processes throughout the Galaxy, which bombard the Earth from all

directions) interact with the Earth's upper atmosphere. Trees absorb carbon-14 during photosynthesis, and traces of it survive in tree rings, giving an historical record of the abundance of carbon-14 produced.

When the Sun is active, more charged particles stream out from the Sun into the solar wind. They travel through the Solar System, some encountering the Earth's magnetic field on the way. As the particles do not easily cross magnetic field lines, they move around the Earth, creating a barrier to cosmic rays. Thus the higher the Sun's activity, the better the barrier and the less carbon-14 is produced for trees to breathe in. During times of low solar activity, it follows that more carbon-14 is produced. Maunder's results were corroborated because the carbon-14 content in tree rings is abnormally high during the period 1645 to 1715.

Another, less well-defined minimum, called the Spörer minimum after the German astronomer Gustav Spörer, occurred between 1450 and 1540. There was also a period of unusually high activity, the Grand Maximum, between 1000 and 1250.

photosphere. There, temperatures are about 300 K greater than in the surrounding photosphere. Faculae appear close to sunspots all across the solar disk, but are generally only seen at the limbs of the Sun (the edges of its visible disk). This is because any faculae on the disk itself are viewed against the brighter, hotter solar material beneath. At the limbs, faculae are seen against the cooler, outer layers of the Sun, giving greater contrast. The Sun's limbs appear darker than the disk, again the effect of looking through material in the center of the solar disk through to the hotter, brighter internal regions. This contrast in brightness is known as limb darkening.

In addition to the faculae associated with sunspots, polar faculae can sometimes be observed at latitudes higher than 55° north and south, where sunspots are rarely seen. Like the faculae associated with sunspots, polar faculae appear more numerous during the rise to solar maximum than at solar maximum itself.

The solar spectrum

Since the solar material is a plasma, electrons are moving freely between the different energy levels (or orbits) in the hydrogen and helium atoms as they gain and lose energy (see the feature on pages 10–11, "Inside the atom"). When the German astronomer Josef von Fraunhofer examined the spectrum of the Sun in 1814, he discovered it was crossed by many dark lines, now known as Fraunhofer lines (see page 31). Today it is understood that these lines are produced by electrons moving from one energy level to another and absorbing photons of a particular energy, leaving the dark (absorption) line in the spectrum. When the electron drops back down to a lower energy level, it will emit a photon and create a bright (emission) line.

By studying the patterns of absorption and emission lines, solar scientists can identify the chemical composition of the Sun, and also determine densities, temperatures and wind speeds. These spectral lines are critical tools for the detailed exploration of objects from a distance, and simple spectroscopes can be constructed and used by amateurs. It is also possible to construct spectrohelioscopes for observing the

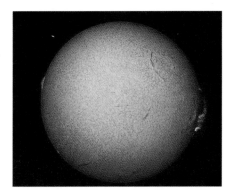

▲ *A prominence can be seen in this hydrogen-alpha image of the solar disk. The dark lines across the disk are filaments.*

Sun at one selected wavelength, revealing different features depending on the wavelength selected (see Chapter 2).

The Sun in hydrogen alpha

Hydrogen alpha (H-alpha or Hα) is a strong Fraunhofer line. It is the first line in a series of lines known as the Balmer series, which are found in the red part of the spectrum. The series is named for the Swiss astronomer Johann Balmer, who studied the lines in 1885. The H-alpha line represents the transition of an electron from the third to the second energy level in the hydrogen atom. The radiation emitted by the electron moving between these levels has a wavelength of 656.3 nanometers (nm, 1 nm being equivalent to 10^{-9} m) or, in the traditional angstrom units, 6563 Å (1 Å being equivalent to 10^{-10} m). Filters are available for observing only in this wavelength (see Chapter 2).

Prominences, plages and spicules

Observing with an H-alpha filter reveals a layer of the Sun up to 10,000 km above the photosphere: the chromosphere. With an H-alpha filter, features such as prominences can be observed. Prominences are huge outpourings of solar material seen at the limb of the Sun. The solar material is contained within magnetic field lines, so prominences often appear as relatively thin loops, arching many thousands of kilometers above the limb of the Sun.

There are two kinds of prominence: eruptive and quiescent. Quiescent prominences can remain suspended above the Sun for months, showing little change, while eruptive prominences change over shorter timescales, often fluctuating in distance from the solar limb. Sometimes the magnetic field lines break and the solar material escapes from the Sun entirely. Prominences seen against the solar disk appear as dark filaments. They can stretch hundred of thousands of kilometers across the solar disk.

Plages are brighter, hotter clouds of chromospheric material associated with sunspots. Their white-light counterparts are faculae. Also visible in H-alpha are spicules – flame-like columns of solar material that rise about 10,000 km above the photosphere, along magnetic field lines. Spicules have a similar lifetime to granules, of around ten minutes.

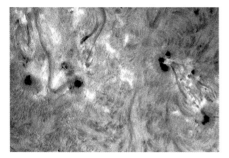

▲ The dark lines are prominences seen against the solar disk.

2 · INSTRUMENTS FOR THE OBSERVER

It is easy to observe the Sun in ordinary, white light. You do not need complicated or expensive equipment, and it is fun to follow solar activity by observing sunspot groups (see Chapter 4). But sunspots are just one manifestation of solar activity (see Chapter 1). Closely associated with sunspots are changes in the solar magnetic field, and the appearance of flares, prominences, coronal holes and coronal mass ejections. For decades, professional astronomers have been observing the Sun in other wavelengths of the electromagnetic spectrum and with specialist instruments which reveal different aspects of solar activity. As with many other branches of astronomy, recent technological advances have brought non-white-light observing into the realm of the amateur.

Observing the Sun presents the astronomer with the exact opposite of the problem met when observing other celestial targets: instead of trying to pick up as many photons as possible from very faint objects, we have to minimize the light and energy coming from the source. Safety is also an aspect because looking directly at the Sun, even without any optical aid at all, can seriously damage your eyes and cause blindness. Never look directly at the Sun or underestimate its power. Taking just a few sensible precautions means solar observing can be done easily with safety, and the rewards are many. (See page 36, "Safety measures.")

Equipment for white-light observing

One of the benefits of observing the largest and brightest object in the sky is that it is not necessary to have sophisticated and expensive equipment. It is easy to project an image of the solar disk on to white card with a pair of binoculars, allowing sunspots to be seen (see Chapter 3). Even without binoculars, a simple pinhole camera can be made, allowing observation of the larger sunspots (see the feature on pages 24–25, "Pinhole cameras").

Telescopes

Unusually for astronomy, the smaller and cheaper telescopes are best for solar observing. Larger telescopes should have their apertures reduced, otherwise too much heat will travel through the tube into the eyepiece. This "stopping down" can be done fairly easily by fixing a lens cap over the objective end of the telescope (the end nearest the Sun). A circular hole of the required dimension (no greater than 30 mm) is cut into the lens cap, offset from the center to further reduce the build-up of heat. This cap can be made simply out of strong black card, but ensure it is fixed securely before turning the telescope to the Sun.

Both refractors and reflectors can be used for solar observing, although refractors are usually preferred. Refractors are telescopes that use a lens to focus the light from the object being observed, while reflectors use a mirror to collect and focus the light from the object. In a Newtonian reflecting telescope, the light hits the mirror and is reflected back up the telescope to a small, optically flat mirror. This mirror, called the secondary mirror, then deflects the light sideways to one side of the tube or frame, where the image can be viewed in the eyepiece.

Today there are telescopes that use a combination of lenses and mirrors, for example Cassegrain, Schmidt–Cassegrain and

Pinhole cameras

One method of observing the Sun that needs no equipment at all is to use a pinhole camera. This can be created simply by making a hole in a piece of card and projecting the Sun's image through the hole on to a piece of white paper. A small but perfectly focused image of the Sun can be obtained in this way, and depending on the size of the hole and the distance at which the Sun's image is focused, some of the larger sunspots can be seen.

Do not make the hole too large: start off with – literally – a pinhole. The larger the hole, the larger the projected image, but the farther you will have to move your projection card to obtain a sharp image. It is more difficult to obtain a sharp image with a large hole.

The more protection from the Sun's direct rays you can give your projected image, the easier it will be to see any sunspots. You can use a cardboard box to help screen the image from the surrounding sunlight.

A larger image using this "pinhole" technique can be obtained if the image of the Sun is projected across a darkened room. One of the first scientific solar observations was made using this so-called camera obscura technique by Gemma Frisius on January 24, 1544, when he watched a solar eclipse projected through a hole in a wall. A famous illustration of this event was published in De Radio Astronomica et Geometrico in 1545 and has subsequently figured in many astronomical texts. There is no need to make a hole in your wall – this effect can be obtained by using blinds or even black paper taped over a window. Gemma Frisius and other early solar observers used this technique to great effect.

One calculation that can be done with observations with a simple pinhole image is to estimate the diameter of the Sun. The ratio of the size of image produced to the actual diameter of the Sun is the same as the ratio of the distance of the image from the pinhole to the distance between the Earth and the Sun, as follows:

Maksutov. Observing the Sun with these is not recommended. These telescopes are sealed, and the build-up of heat inside them could damage their optics. Some have plastic components located in the path of the Sun's rays which can easily melt in the heat, ruining the telescope. The more basic the type of telescope, the better it is for solar observation.

Reflectors

Reflecting telescopes use mirrors made of glass coated with a thin layer of aluminum. Simple reflectors can be used to observe the Sun, but

$$\frac{\text{diameter of image}}{\text{diameter of Sun}} = \frac{\text{distance of image from pinhole}}{\text{distance of Sun from Earth}}$$

So, knowing the Sun–Earth distance (1.496×10^8 km) and measuring the diameter of the hole and the distance from the image to the pinhole (ensure they are measured in the same units, e.g. both in millimeters or both in centimeters), you can estimate the diameter of the Sun:

$$\text{diameter of Sun} = \frac{\text{distance of Sun from Earth x diameter of image}}{\text{distance of image from pinhole}}$$

The Sun's diameter in kilometers is then:

$$\text{diameter of Sun} = \frac{1.496 \times 10^8 \text{ x diameter of image}}{\text{distance of image from pinhole}}$$

The actual diameter of the Sun is 1.39×10^6 km.

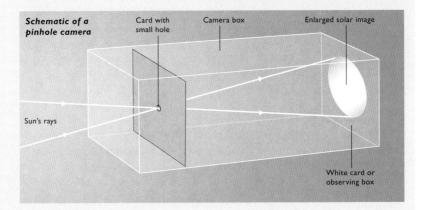

Schematic of a pinhole camera

Card with small hole

Camera box

Enlarged solar image

Sun's rays

White card or observing box

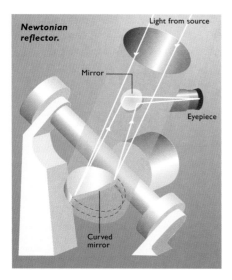

Newtonian reflector.

Light from source

Mirror

Eyepiece

Curved mirror

◀ *In a Newtonian reflector, light from the source travels down the tube of the telescope to a curved mirror which focuses the image via a secondary flat mirror into an eyepiece at the side.*

▼ *A refracting telescope uses lenses to focus the image of the source at an eyepiece.*

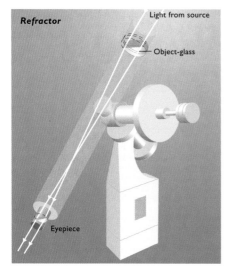

Refractor

Light from source

Object-glass

Eyepiece

because they are generally open at the objective end, they allow the heated air in the tube to exchange with the cooler air outside, creating turbulence and making it difficult to focus the image.

Another problem with reflectors is that the secondary mirror is at the focal point (the focus of the rays), so can become very hot. This can distort the mirror and hence the image.

The optimum aperture for a solar observing reflecting telescope is between 150 and 200 mm. Anything larger must be stopped down as described above.

Refractors

Refractors are the archetypal astronomical telescope, with a lens at the objective end (the end nearest the sky) which focuses the light from the object being observed into another lens (the eyepiece) at the other end of the tube. As the tube is enclosed, the air within the tube does not become so hot, thereby minimizing the amount of air turbulence, which distorts the image.

Ordinary terrestrial telescopes, such as those sold for birdwatching, can be used for solar observing, but it is better to use a telescope designed for astronomical use. In general a terrestrial telescope will have a combination of lenses within it to allow the image of the object under observation to be viewed the right way up. In astronomical refracting telescopes, unless other lenses or mirrors are used, the image is upside down. Each time another lens is introduced into the system, it darkens the image slightly because the glass from which the lens is made absorbs some light. Also, if any cement is used between lenses, the heat from the Sun can melt it.

The optimum aperture for a solar observing refracting telescope is between 60 and 100 mm. Anything larger must be stopped down as described above.

Telescope mounts

Any telescope must be on a firm mount, the firmer the better as any movement of the telescope is amplified in the image. It is very difficult to observe an image which is shaking.

As with the telescope itself, for solar observing it is not necessary to have the most expensive type of telescope mount. A simple, relatively cheap refractor bought at a high street shop will normally come with the simplest type of mount – an altazimuth. This mount allows the telescope to be moved about two axes: horizontally and vertically. The Sun travels around the sky at an angle to the horizon, so to follow the Sun you have to move the telescope in two directions every couple of minutes. With practice it becomes easy to follow the Sun's motion with a combination of horizontal and vertical movements, though your method of projecting the Sun's image will affect the degree of difficulty (see Chapter 3).

A more sophisticated (and therefore more expensive) type of mount is the equatorial. This mount has one axis which has to be lined up with the north (or south) celestial pole. This point is a projection of the Earth's own north or south pole on to the sky and is the point around which the rest of the sky appears to rotate due to the rotation of the Earth. Once one axis of an equatorially mounted telescope is lined up with the celestial pole, the other axis allows the telescope to move in the direction that directly follows the movement of celestial objects. Thus for this type of mount, movement in one axis only is needed to follow the Sun.

More expensive equatorial mounts have a slow motion drive which automatically moves the telescope at the same rate as the Earth's rotation, but in the opposite direction, allowing the user to observe an object without having to move the telescope manually at all.

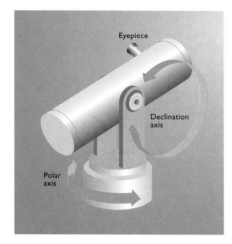

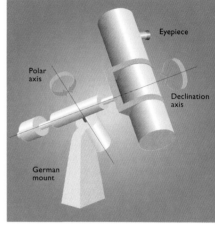

▲ The altazimuth mounting is the simplest form of astronomical mount, giving movement in two axes. With both horizontal and vertical movement, the telescope can follow the object as the Earth turns.

▲ The equatorial mounting is a mount that is aligned to the Earth's celestial north or south pole. In this manner the telescope is carried around by the Earth's rotation and only has to be moved in one axis to follow a celestial object.

Eyepieces

High-magnification eyepieces are not needed for solar observing. Also avoid any type of eyepiece made of two or more lenses that are cemented together at their inner surfaces because the heat from the Sun can destroy the cement, ruining the eyepiece. Use the cheaper eyepieces – types such as Huygens, Mittenzwey or Ramsden – as they are of basic design with no internal cementing.

White-light filters

Safe solar filters allow the Sun to be viewed directly, but you must be very, very sure that the filter you intend to use is safe. If you are not absolutely certain, do not use it. To be safe, a filter must reduce the amount of light reaching the eye by a factor of 100,000 or more (these are filters with an optical density of 5 or higher). A cheap alternative is welder's glass of number 14 or higher, but although useful for naked-eye observing, optically it is not generally very good.

Some telescopes are sold with a solar filter included. Treat these filters with extreme caution – many are not safe. My advice would be to throw them away and either always observe the Sun by projection, or invest in a proper filter from a reputable source.

Make sure that any filter you use has been designed for the purpose you wish to use it. While there are some filters that are perfectly safe for naked-eye viewing, they would not be safe for use with magnification.

Objective filters

Filters that attach to the observer's end of a telescope should be avoided as the Sun's rays are then focused on the filter, heating it to extreme levels and possibly causing it to shatter while in use. It is always best to use an objective lens filter – one that fits over the end of the telescope pointing at the Sun. An objective lens filter basically consist of a partially transparent mirror coating on glass or polyester. The mirror coating reflects most of the light and heat away, allowing a safe proportion of the Sun's light to enter the telescope.

Glass objective filters are usually coated with chromium or Inconel (a nickel–chromium alloy). There are usually few problems with their use, but the glass base has to be ground and polished very precisely to avoid any distortion in the image, which makes these filters quite expensive. They must also be handled very carefully as any damage will compromise their safety. Many observers prefer glass filters to polyester filters as they make the Sun's image appear a natural yellow-orange color. A reputable manufacturer of glass filters is Thousand Oaks Optical in California, USA.

Polyester objective filters are usually coated with aluminum. These are less expensive than the glass equivalent, but there are large variations in the standard of the polyester film base. The tradename for this type of material is Mylar, but some material sold as Mylar is not of sufficiently good quality to filter the Sun's light properly. Even some eclipse glasses sold expressly for the purpose of directly viewing the Sun have used this poorer quality Mylar. To be safe, the Mylar film must be aluminized on both sides and have no scratch marks or pinholes. Always buy from a reputable dealer, and remember that the price reflects quality.

Mylar film produces a bluish solar image. Mylar-based filters have to be handled with care as they can easily become worn or scratched. It only takes a tiny amount of unfiltered solar light to cause blindness, so always carefully check your filters for wear before using. Unless mounted stiffly, the polyester filter will flex with the wind, causing the image to distort.

Other materials and filter that are sometimes claimed to be suitable for solar observation, but are *not* safe, include smoked glass, exposed photographic film, photographic polarizers and neutral density filters. In general, unless you are absolutely certain that the filter you have is safe, undamaged and from a reputable source, I would advise observing the Sun by projection only.

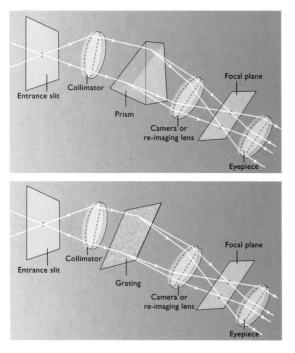

◄ In a spectroscope with a prism, the prism breaks the light from the object into its constituent parts, which are spread out and viewed at the eyepiece.

◄ This spectroscope uses a grating instead of a prism, but the principle is the same as above.

Spectroscopes

The Sun's spectrum reveals a great deal of information about its chemical composition (see Chapter 1). Spectroscopes can now be purchased to fit on amateur telescopes, or they can be fairly easily constructed, and many amateurs can now display the Sun's spectrum with a resolution sufficient to observe the Fraunhofer lines.

A spectroscope consists of an entrance slit, a collimator, a prism or grating, and a system for displaying or recording the spectrum. Sunlight entering the slit is converged to form a parallel beam by a system of lenses or mirrors called a collimator. This parallel beam is then spread out into a spectrum by either a prism or a grating, and the result can be viewed optically or – as is more usually the case today – imaged or photographed. An instrument that records a spectrum is known as a spectrograph.

Both a prism and a grating will spread the incoming light into a spectrum by dispersing the constituent wavelengths by different amounts. The amount of dispersion caused by a prism depends on the material of which it is made: the better the material, the more dispersion, but also the more expensive. For this reason, many amateurs use diffraction gratings.

An optical diffraction grating consists of a plate on to which grooves or lines (typically 1000–1500 per millimeter) are inscribed. If the plate is made of glass, it is known as a transmission grating; if of metal, as a reflection grating. The closer the lines on the grating, the more the light is spread out.

Solar composition

Once a spectrum is obtained, it can be photographed or imaged digitally. If the resolution is high enough for Fraunhofer lines to be seen, individual lines can be identified, revealing the composition of the solar material under observation. Some purpose-built spectroscopes are calibrated with a scale that is projected on to the spectrum, allowing the different lines to be identified easily. For uncalibrated spectroscopes, if a line of known wavelength can be identified (for example, the strong lines in hydrogen or calcium), the wavelength of others can be worked out. The lines can then be identified from a table of the wavelengths of characteristic lines in the spectra of chemical elements.

It is interesting to compare the spectrum of a large sunspot with that of the photosphere. Since the material within the sunspot can be 2000 K cooler, the lines will differ quite considerably from those in the photospheric spectrum.

Spectrohelioscopes

Whereas a spectroscope shows the whole of the Sun's spectrum, a spectrohelioscope gives a view of the Sun in the wavelength of just one spectral line. As with a spectroscope, the Sun's light is dispersed by a prism or diffraction grating, then a second slit is used to view a precise wavelength. If the Sun's disk is scanned rapidly, a record of the entire solar disk can be built up in the selected wavelength. Such images are called spectroheliograms and are produced by spectroheliographs.

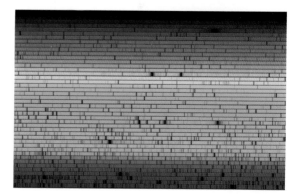

▶ Fraunhofer lines are dark absorption lines viewed across the spectrum of the Sun. First seen by Fraunhofer, the lines are produced when electrons absorb energy when changing to lower energy levels.

Observing the Sun in a particular wavelength reveals different solar layers and features since each wavelength corresponds to a different temperature. For observing visually, the wavelength chosen has to correspond to a wavelength to which the eye responds, but photographically this range can be extended into the ultraviolet (for example using the calcium II, H and K lines) and the near infrared. One line used to reveal features in the chromosphere is the first line in the Balmer series, the hydrogen-alpha line (see Chapter 1). Today there are spectrohelioscopes specifically built for use by amateurs in the hydrogen-alpha (H-alpha) wavelength, known as H-alpha filters.

Hydrogen-alpha filters

As explained in Chapter 1, hydrogen alpha (often abbreviated to H-alpha or Hα) is the first line in the Balmer series and corresponds to a wavelength of 656.3 nm. In H-alpha radiation some of the more dramatic results of solar activity are visible because this wavelength corresponds to a temperature of around 10,000 K – a temperature that occurs within the chromosphere, from where material is emitted in

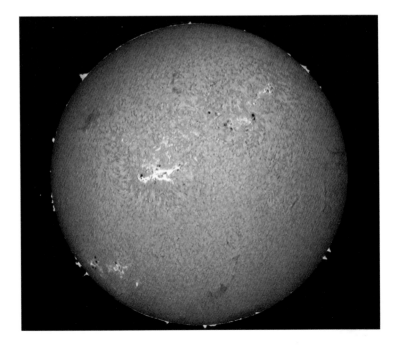

▲ The Sun in H-alpha shows features in the chromosphere such as flares, prominences, filaments and spicules. The sunspots can also be seen.

▶ *Many affordable telescopes devoted to the study of the Sun in H-alpha are now on the market.*

prominences and some flares. Luckily for us, H-alpha radiation lies in the visible part of the electromagnetic spectrum, so it arrives at the ground, is not absorbed by the Earth's atmosphere and can be seen by our eyes.

Today there are many filters on the market which, for a moderate outlay, will allow anyone with a telescope to watch the Sun in H-alpha. Some observe the Sun in only one or two "fixed" bandwidths, while others allow the observer to tune the filter, allowing observations to be made at a choice of bandwidth. Different chromospheric effects can be seen depending on the bandwidth selected: the narrower the H-alpha line, the more clearly the chromospheric effects become.

Seeing red

Viewed in this narrow bandwidth, the Sun is breathtaking. Even the most jaded amateur cannot fail to be awed by the sight of a seething red solar disk with huge outpourings of solar plasma – prominences – spewing out from the limb. Material being ejected out toward the Earth is visible as dark filaments meandering across the disk. Although it is not possible to watch the Sun's material move in real time – the distances are just too great – it is possible to notice a change in the shape of the prominences and filaments over a time span of just an hour, and when watching the Sun through an H-alpha filter it is quite possible to lose an hour or so of time! There are few more impressive sights than our nearest star in eruption.

Observing the Sun in H-alpha has the same broad appeal as white-light observing. You may just wish to look occasionally to see what is going on or you may wish to undertake a more serious observing program. Obviously, the type of filter that you have will determine what you can observe, but there is an enormous amount of data to study and collect, from just counting the number of prominences and filaments to classifying them and measuring their frequency and even measuring their velocities. There is also some important science to be done, by associating flares and other activity to particular sunspot areas. You do not need an H-alpha filter to do this if you have access to the Internet, because you can use the images published daily from professional observatories such as the SOHO spacecraft (see Chapter 9).

The right choice

Choosing the right filter will depend upon your interest. There are two basic types of filter which allow you to observe chromospheric effects: the prominence telescope and the Lyot filter. The prominence telescope, as its name suggests, only allows observation of phenomena at the limb of the Sun, while the Lyot filter allows observation of both limb phenomena and surface chromospheric effects across the solar disk.

The prominence telescope

Instruments that reveal solar activity at the limb can be bought by the amateur observer. They are available as entire telescopes or as attachments. They work on the same principle as the coronagraph, an instrument that produces an artificial eclipse by introducing an occulting disk into the field of view that blocks out the majority of the photosphere, allowing just the solar limbs to be observed. The use of a filter then reveals the chromospheric limb features such as prominences, plages and spicules (see Chapter 1).

▼ *Prominences are huge eruptions seen at the limb of the Sun in H-alpha filters.*   *In this image, the Earth to scale is shown in the right-hand corner.*

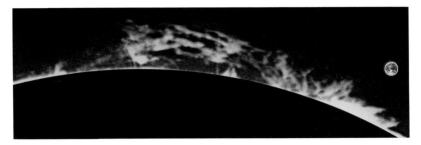

The Lyot filter

Today, reasonably priced solar telescopes are available based on the Lyot filter (a polarizing interference filter). They allow the whole of the solar disk and limb phenomena to be observed simultaneously. Unless your interest lies in the construction of your own equipment, these telescopes have largely superseded home-constructed spectrohelioscopes. In addition to the limb features visible with a prominence telescope or attachment, the Lyot filter will show filaments – the dark lines of prominences seen against the disk. Solar flares can also be observed.

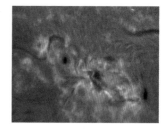

▲ *Dark lines viewed across the solar disk in H-alpha filters are prominences directed toward the Earth.*

Listening to the Sun

In addition to radiation in the visible part of the electromagnetic spectrum, radio waves emitted by the Sun can be detected at the Earth's surface. Radio astronomy began in the 1930s when the amateur astronomer Gröte Reber built the first radio telescope, but it was not realized until World War II that the Sun is a powerful emitter of radio waves. In 1942, British army radar operators picked up interference on their equipment. Originally it was believed the Germans had discovered their use of radar and were jamming the signal, but eventually it was realized that the radio interference came from the Sun. Solar radio emission is very closely related to solar activity. In 1942, a large sunspot group had been crossing the solar disk. Associated with this group were large solar flares, and it was the radio emission from the flares that had jammed the British radar.

Radio emission from flares is generally in the meter wavelength range of the electromagnetic spectrum and is very sporadic, corresponding to the release of energy within large sunspot groups. There is also a more constant component of solar radio emission in the centimeter wavelength range, found to be emitted by the chromosphere, with some of the shorter wavelengths emitted from the corona.

Amateur astronomers can build radio telescopes with antennae tuned to pick up emissions from the Sun, but a detailed discussion of this equipment is beyond the scope of this book.

——— 3 · HOW TO OBSERVE THE SUN———

Some amateur astronomers look upon solar observers as some kind of poor cousin: they do not have to know their way around the stars to pinpoint a particular planet, they do not have to stay up all night in the freezing cold to view that faint comet, they do not need the largest telescope possible to be able to capture every photon of a remote galaxy. For these detractors there is no astronomical challenge in observing the Sun. They could not be more wrong!

As the brightest object in the sky, the Sun should be the easiest object to observe, but this is not the case. The brightness of the Sun creates its own problems. Granted, you do not need to know your way around the stars to find it, but surprisingly it can be difficult to focus your telescope on the Sun. It is usually warmer when the Sun is out, but it can still be very cold, and it is true that you do not need a huge telescope to capture every photon – quite the reverse – but the Sun's intensity has to be dealt with appropriately.

Once you start to observe the Sun, you quickly realize that although solar features can be observed with a minimum of fuss and equipment, to study them in more detail is not so easy. The rapid changes visible on the solar disk over a short period of time (sometimes just minutes) produces a fascination not available with other astronomical targets – observing the Sun can become quite addictive. If you decide to record your observations, it is easy to contribute data to the professionals, but delving deeper into the data by your own analysis reveals hidden complications. Solar observing offers enjoyment to observers at all levels.

As discussed earlier, it is not necessary to have a large, powerful telescope to observe the Sun. Indeed if you do have a large telescope, you must stop it down to reduce the amount of light and heat captured in the system or you will damage the instrument (see Chapter 2 on how to reduce the aperture of a large telescope.)

Safety measures

- **Never** look directly at the Sun – either with or without magnification – unless using appropriate and undamaged filters.
- **Never** squint along the tube of a telescope at the Sun.
- **Never** leave a telescope or binoculars pointed at the Sun unattended.
- **Never** use filters unless you know they are the appropriate ones to use, they are from a reputable dealer and are in good condition.
- **Always** supervise casual observers and children near instruments pointing at the Sun at all times.
- **Always** ensure that the finder of a telescope used to observe the Sun has its lens cap on at all times.

You do not even need a telescope: the Sun can be observed with binoculars, as described on pages 44–45.

Using a refractor: getting started

The best instrument to use is a small refracting telescope with an aperture of about 60 to 100 mm. The first task is to point the telescope at the Sun. This may sound obvious and easy, but in fact it is quite a difficult thing to do.

You cannot use the telescope's finder in the usual way as you will blind yourself. For safety, the finder should always have its lens cap on when the telescope is being used to observe the Sun. This is because when the telescope is pointing at the Sun, the finder (assuming it is accurately aligned) will also be pointing at the Sun. If you accidentally move your eyes through the view finder's line of vision, you can damage your eyes. The view finder will also be focusing the Sun's rays, and if you stand behind the telescope the focused rays can burn you or set light to anything flammable.

You should not even glance along the telescope tube to check that it is pointing at the Sun. You can damage your eyes if you are just looking at the Sun for a few seconds, and if you make a habit of setting up your telescope for solar observing by doing this, the damage can accumulate. Make sure that you understand the dangers of observing the Sun. Develop a safe method of observing the Sun from the beginning, never be tempted to take risks and always ensure that anyone else observing with you – especially if a beginner – is aware of the dangers and is protected against them.

▲ Move the telescope until the shadow on the ground is the smallest possible and the telescope will be pointing at the Sun.

The shortest shadow

The safest way to make sure that your telescope is pointing directly at the Sun is to use the shadow cast by the instrument on the ground. As the telescope is tube-shaped and presents its smallest profile to the Sun when aligned for solar observation, that is also when it produces its smallest shadow.

With the lens cap of the telescope still on, point the telescope roughly in the direction of the Sun – it does not matter how roughly – and resist the temptation to look at the Sun yourself. Then stand with your back to the Sun and concentrate on the shadow. Move the telescope in all directions until the shadow is the smallest in every dimension. This can take a while to do the first time, but practice will allow you to achieve this in seconds.

Then remove the lens cap of the telescope. Hold a piece of paper or white card up at the eyepiece. It does not matter at this point whether the eyepiece has been focused. If the telescope is pointing directly at the Sun, the Sun's image should appear as a bright white disk on the card. If it does not appear immediately, tap the telescope very gently, moving it only small amounts until the Sun's image appears.

The technique of using the shortest shadow to obtain the Sun's image through the telescope optics can take some time to perfect, but it is worth persevering as the method is the safest one to adopt. It can be used with all types of telescopes and with binoculars.

Focusing the image

Once you have the image shining through the telescope, you will need to bring it to a focus. This can be done by getting the image on to a piece of white card near the eyepiece, and gradually moving the card away from the telescope. The farther

◀ Shielding the image from the Sun's direct rays can be done simply by fixing a piece of cardboard over the objective end of the telescope.

away you move the card, the larger the Sun's image becomes. When you have it at an appropriate diameter (a 60 mm aperture refracting telescope will produce a good focused solar image of diameter 150 mm), stop moving the card and use the eyepiece in the usual way to sharpen the image. You may be surprised at how far away the card has to be held and how far out the eyepiece has to be moved to obtain a sharp image.

Use the edge of the Sun for focusing. The sharper the solar circle, the more the image is in focus, though you may need to alter the focus slightly when concentrating on solar details. The granulation, for example, will be in focus for a slightly different configuration than for sunspots.

If you draw a circle on the card at an appropriate diameter, and hold the card so that the Sun's image fits exactly inside the circle, you will know that you are holding the card exactly perpendicular to the telescope. If the Sun's image is not circular, you will need to adjust the angle of the card.

A dark image

Holding white card behind the eyepiece will allow you to view the Sun's image, but this image will also be illuminated directly by the Sun's rays. This will wash out the contrast of the image produced by the telescope, making it hard to see detail on the solar disk. To prevent this from happening, the card should be shielded from direct sunlight as much as possible. The darker the surroundings of the card, the more detail will be visible on the solar disk.

▶ The more the image is protected from the Sun's direct rays, the better the image. This again can be achieved simply by the use of a cardboard box.

The easiest way to shield the image card is to use another piece of card at the end of the telescope nearest the Sun – the objective end. Make a hole in this second piece of card and fit the card over the objective end. Ensure that the card is large enough to prevent the direct rays of the Sun from falling on to the image card.

Another way to do this is to use a star diagonal on the telescope. Star diagonals are usually supplied with refracting telescopes as they make observing easier when the telescope is positioned to observe something at an awkward angle. A star diagonal contains a small flat mirror set at an angle of 45° and fits into the eyepiece socket of the telescope in front of the eyepiece. The image is thus bent around a 90° angle, allowing it to be viewed from the side of the telescope.

For solar observing, the image can then be projected sideways into a box. The sides of the box shield the image card from the direct rays of the Sun.

An observing box

If you intend to observe the Sun frequently, you may like to construct a more permanent observing box. Some people have a dedicated telescope for solar observing – usually a small refractor which would not otherwise be used. For a telescope that is to be used only for solar observing, an observing box can be permanently fixed at the correct distance from the eyepiece, thereby removing the necessity of focusing the image at each observing session.

It is possible to make an observing box that is removable to enable the telescope to be used for observing objects other than the Sun, but that will require a certain level of do-it-yourself expertise. However, it is not necessary to build a complicated observing set-up. My system for observing the Sun uses a non-attached cardboard observing box which is easily set up and dismantled. Using this simple system I can make sunspot counts and drawings, but I can also use my telescope for observing other astronomical targets.

However, there are benefits in having a permanently mounted solar observing box: it is easier and quicker to set up, and all your observations are done with the same observing criteria. Obviously, the design of the observing box will depend on the type of telescope you have.

If your telescope is permanently mounted in an observatory, then an observing box can be constructed as part of the observatory instead of as an attachment to the telescope. This overcomes the problem of balance on the telescope – if you build something on to one end of a telescope, it will interfere with the telescope's center of gravity, requiring a counterweight to be added – and also frees the telescope for other observing activities.

To construct the box in the observatory, first decide on the size of image you require, and what size you can achieve with your telescope and within the confines of your observatory. You may like to experiment with what size image you can achieve on the wall of the observatory because if this is satisfactory, you immediately have the place you need to fix your observing box. An image of about 60 mm will be large enough to show the largest sunspots. An image of at least 100 mm is needed to produce data sufficiently accurate to send in for professionals to use, and most observing societies work with images 150 mm across.

If you do not wish to fix the box on to the observatory wall, then you will have to make a structure that fixes to a particular point on the floor. Once you have decided

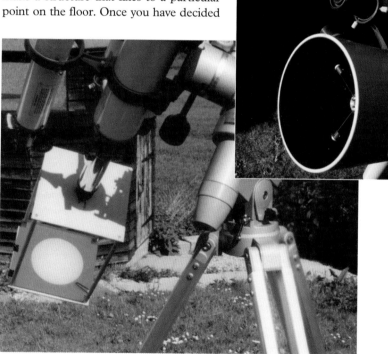

▲ Two examples of observing screens permanently fixed to telescopes. The main picture shows a box attached to a refracting telescope. The side of the box nearest the Sun shields the image projected on to the observing screen. The inset picture shows a reflector with a screen fixed so that the solar image is projected sideways through the eyepiece.

the distance you want the box to be from the telescope, the next problem is to determine at what height the observing box will need to be and to decide how you will achieve this height, either with a temporary structure that can easily be fixed and removed at the observing distance, or with a permanent structure that still allows you to use the telescope for other observing.

Bear in mind that if the observing box is not fixed directly to the telescope, then the height of the observing box will need to be adjusted according to the time of day that observing is undertaken. The Sun can be observed at all times of day, and hence at different angles in the sky.

If you are using a free-standing telescope and wish to fix the box to the telescope, then you will have to work out how far back you need your observing box, and a method of constructing a support from the telescope to the box. The farther the distance from the telescope you want your observing box, the harder it is to support it.

One method of attaching an observing box to a telescope is by two thin lengths of wood (about 20 mm wide and 10 mm thick). They can be attached to the telescope tube using Jubilee clips (hose clips). If the wooden struts are attached at an angle to the telescope, sloping downward or upward away from the line of vision, the observing box can be supported at its base or top by the other ends of the struts, thus allowing the image to be seen cleanly without any obstruction in the way.

The box itself does not need to be constructed out of heavy or expensive material: cheap wood will do. The heavier the box, the more weight will need to be added to the telescope to act as a counterbalance. The back of the box should be smooth enough for the image not to be distorted, and if you intend to make paper records of the Sun it should be possible to fix either a graticule or an observing blank to the back of the box. If you intend to draw your observations directly on to an observing blank, the blank will need to be capable of being fixed and removed easily. Thin plastic sheeting fixed to the box provides a good surface to which to attach a blank for making observations and also allows the use of pins or sticky tape. If pins are used, you may find that you need to replace the sheet at intervals, because the accumulation of pinholes makes an uneven surface for drawing your observations.

An observing box that is attached to a telescope should be as light as possible, but not so light that it will flex in any wind. If sufficient screening from the direct rays of the Sun can be achieved, it does not need to be a box, just a flat observing platform on which you can view the image and fix your paper if required.

▶ *If the image is projected immediately behind the telescope (as is the case when using a refractor without a star diagonal), a screen needs to be fixed to shield the image. Divers' angle weights, which are easy to attach and detach, make suitable counterweights to balance the telescope once the screen is attached.*

If you are using a refractor with the image projected straight out behind the telescope, you will need to shield the observing box from the direct rays of the Sun. This can be done by fixing a screen at the objective end of the telescope that is large enough to cast a shadow over the whole of the image. If the observing box is attached to your telescope, the weight of this screen will help to counterbalance the weight of the box.

If you are using a free-standing observing box not fixed to the telescope, then you will need a very lightweight screen to shield the image, otherwise you will need to construct a counterweight for the screen. To enable you to use your telescope for other observing, the screen should be easily detachable. This can be achieved by making a circular sleeve from thick card which just slips over the objective end and attaching the screen to the sleeve. If necessary, fix small stops to the telescope tube to hold the sleeve in place. The stops can be made from card and left on the telescope permanently. They will not be heavy enough to destroy the balance of the telescope, nor large enough to interfere with other kinds of observing.

A lightweight screen can be made out of thin, but rigid card. It would need to be thick enough not to flex in any wind, as this movement will cause the telescope to move and blur the image. It is possible to buy sheets of thin plastic which can be used as screens.

If the observing box is heavy and fixed to the telescope, and the screen is not heavy enough to counterbalance the box, you will need to attach some extra weights to the telescope. Anything of the correct weight and easily attachable to the telescope will do, but I find divers' ankle weights ideal. They are available in a variety of

Using binoculars

It is very easy to observe the Sun with binoculars, and using binoculars will show you the disk of the Sun and the largest sunspots. Block off one half with some card. This card can also be used as a screen to prevent the projected solar image being washed out by the direct rays of the Sun. One way of doing both these tasks is to cut a hole in the card large enough to fit over one half of the binoculars, but covering the other half. The card should be fixed over the larger end, the end nearest the Sun, and taped securely.

Direction of light from the Sun

Set the binoculars' focus to infinity. If you have a tripod, fix the binoculars to them, and obtain the Sun's image through them by using the shortest shadow method, described on pages 37–38. Never look directly through the binoculars to get the Sun's image through them, as this will cause blindness. Never even glance along them to get an approximate position. Since you are observing the Sun, the Sun will be out and will be creating a shadow of the binoculars. Move the binoculars around both the horizontal and vertical axes until the shadow is as small as possible. This position should correspond to the Sun shining directly through them. This method can take some time to weight sizes, are fairly cheap, and they have a strap which, unless your telescope is of very large aperture, will fix easily around the tube. If you need extra strapping, this too can be bought from a diving shop.

If you use a star diagonal to project the image sideways from the telescope, the side of the observing box itself can be used to shield the image from the direct solar rays. This can also be done if you are using a reflector and projecting the image sideways.

To minimize any reflection within the observing box, paint the inside black.

perfect, but persevere as it gets easier each time you do it.

Once the shadow is as small as you can get it, hold a piece of white card behind the end you would normally look through. If the binoculars are positioned correctly,

▲ *Using binoculars to observe the Sun is similar to using a refracting telescope. Once one half of the binoculars has been blocked off, the image will be projected out of the back, as shown above, and will need to be screened from the direct rays of the Sun.*

you should see a bright disk of light: the Sun's image. If you do not see this on the card, jiggle the binoculars gently until you do. You should not need to move the binoculars very far if you have managed to get the shadow as small as possible. Once you see the bright disk of light, move the card back away from the binoculars until the image becomes sharp. You may be surprised how far back you have to move the card.

If you do not have a tripod, you will need to fashion a support which allows the binoculars to be at an angle so that they can be pointed at the Sun with the image projected quite far behind them. One solution is to use a wall or windowsill, with a pile of books supporting the binoculars.

A pair of 7 × 50 binoculars, where the 7 denotes the magnification and the 50 denotes the aperture in millimeters, will produce a solar image of about 60 mm diameter, large enough to show most sunspots. On days of good seeing, the umbra and penumbra should be discernible for the larger spots.

4 · MAKING OBSERVATIONS

Once you have set up your equipment and successfully captured the Sun's image, what can you expect to see? It is very unusual to see a completely blank solar disk. This would most likely be around solar minimum, though it can happen at other times when any solar activity is on the reverse side of the Sun. If you do see a blank disk and it is not solar minimum, check that the image is focused correctly. Any sunspots present may be rendered invisible if the image is slightly out of focus.

Normally there will be at least a few sunspots to observe, and near solar maximum the solar disk will often display a great many sunspots with some very complex groups. The larger sunspots and the groups will show a dark umbra surrounded by a lighter penumbra, while smaller spots and pores will only have the dark umbra. Occasionally, a sunspot or region will be sufficiently large for you to see filamentary structure within the penumbra; this structure is produced by the magnetic field within the sunspot group.

Bright faculae may be observed at the solar limb. Faculae often appear before sunspots erupt and can linger after spots have died, so these bright features can sometimes be seen with no apparent associated sunspots. They can also be seen surrounding a complex area or an area which will become complex. Faculae are observable only near the solar limb, because the solar disk is too bright in the middle to allow them to be seen. Even though sunspots do not appear at high latitudes, do not forget to look near the solar poles for polar faculae.

Sunspots and groups of spots seen at the limb can appear flattened or foreshortened. This is an optical effect – you are viewing the features

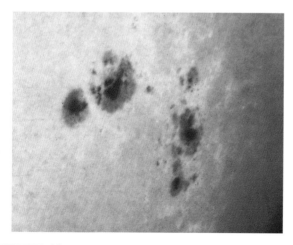

◀ Larger sunspots and sunspot groups can be very complex, with the dark umbral spots surrounded by regions of the lighter penumbra.

at an oblique angle as they appear or recede around the curved edge of the Sun.

If the seeing is very good, the granulation may be observed as a faint dappling over the whole of the solar disk. This effect, along with other small features such as tiny pores, can be glimpsed more easily if the image is moved very slightly by, for example, tapping the telescope gently. It is good practice to get into the habit of tapping the telescope gently once you have looked at all the larger features on view. This may well show up smaller spots or smaller features you will not have noticed. The longer you look at the solar disk, the more chance you have of seeing the small details.

Keeping a record

As with any astronomical observing, it is rewarding to keep a record of what you have seen. In contrast to, for example, deep sky observing, where you are viewing a generally unchanging object, in solar astronomy the subject changes daily. The Sun rotates, and sunspots move across the Sun in a couple of weeks. Sunspots appear, grow, shrink and disappear, and sunspot groups can change their shape over a period of just a few hours.

The changing face of the Sun makes keeping an observing log of the Sun a slightly different project from, for example, recording observations of distant galaxies. The simplest way of noting the activity of the Sun is to count the number of sunspots or groups of sunspots and record whether faculae or granulation are visible. Some observers have their own way of doing this, but the accepted method of counting sunspots is to use a simple empirical formula known as the relative sunspot number (see Chapter 5).

It is very satisfying to keep a log of the relative sunspot number, as this can change significantly from day to day, and if you keep the log for the whole 11-year period of a solar cycle the difference between solar maximum and minimum becomes very apparent. You will also be able to notice shorter periods of activity. Sometimes some regions of the Sun are more active than others. These are known as active longitudes. If there are active longitudes, solar activity and hence the relative sunspot number will fluctuate over a period of about a month: when an active longitude is on the visible disk, solar activity is high, and when the active longitude rotates round to the hidden side of the Sun, the activity can drop quite significantly.

However, the sunspot number alone will not remind you of the shape and spread of sunspots, and it is even more satisfying to look back at drawings of the solar disk. Drawings can be made quite easily, though the method will depend on your observational set-up.

Example of an observing blank

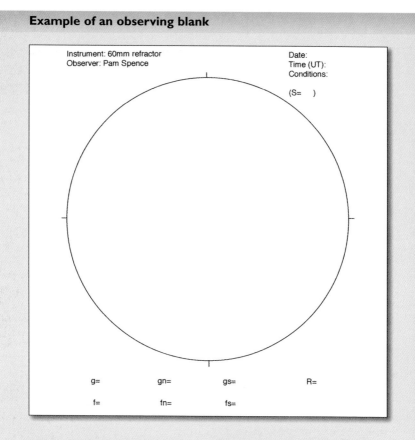

Instrument: 60mm refractor
Observer: Pam Spence

Date:
Time (UT):
Conditions:

(S=   )

g=   gn=   gs=   R=

f=   fn=   fs=

I keep my observing blank simple. Using a 60 mm refractor, I can easily achieve a 150 mm diameter image. I thus have a 150 mm circle with small marks showing the position of the north, south, east and west points. Since I use a refractor with a star diagonal, the image projected into my observing box has the same aspect as the Sun in the sky, with north at the top and east to the left. If you project the image directly out from a refractor, the image will be reversed left–right. Familiarize yourself with how your telescope set-up changes the orientation of the solar disk.

My blank already contains my name and details of my telescope. I then select to record the date, the time in UT, weather conditions, seeing conditions – on a scale from $s = 1$ (excellent) to $s = 6$ (very poor) – and a box of values to allow me to work out the relative sunspot number (R), which will be explained in Chapter 5. On the blank I denote the total number of groups by g, the total number of spots by f, and the proportions in the northern and southern solar hemispheres by g_n, f_n and g_s, f_s respectively.

Other information you can include is your location (latitude and longitude) and the data for the Sun's changing aspect in the sky due to the Earth's orbit (the position angle, the heliographic latitude and longitude of the center of the disk, and the Sun's diameter). For an explanation of these quantities, see Chapter 5.

Once my equipment is set up, I project the Sun's image directly into my cardboard observing box via a star diagonal. This means I can pin my observing blank in the box and draw straight on to the paper. I vary the distance of the box from the eyepiece until the Sun's image exactly fits into the 150 mm diameter circle already drawn. Fitting the image to the circle also ensures that the image is projected on to a flat surface – if the box is tilted at all, the image will not be precisely circular.

To ensure that the north point of my observing blank is correctly aligned, I first draw a faint pencil diameter line across the middle of the paper observing blank. I then twist the paper

until the motion of the Sun across the box takes a sunspot along the line. This indicates that the observing blank is tilted at the correct angle to take account of the inclination of the Sun's axis of rotation in the sky at the time of observing. Once I have done my drawing, I rub out the penciled-in diameter line as it is not needed for the final drawing. My observing blank is shown here at a reduced size. Photocopy it at 200% for a standard solar image of diameter 150 mm.

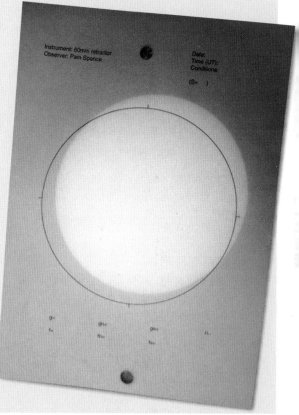

An observing blank

If you plan to draw the solar disk regularly, it is a good idea to design an observing blank which will allow you to make standardized drawings. Having your drawings in a standard format will make it easier to compare them. It will also take less time to draw the Sun if you already have a preprinted form.

Various organizations collate observations from amateurs and forward data to the professionals. If you intend to submit your drawings to one of these organizations, note that each will have its own criteria for recording and submitting observations. They may have their own standard blank, and at the very least they will specify the diameter of the solar disk they require. Some of these organizations are listed in the bibliography.

To a certain extent, the size of telescope you have will dictate the diameter of solar disk you can produce. Most organizations stipulate that drawings should show a solar disk of diameter 150 mm. This diameter of solar disk can easily be achieved using refracting telescopes of 60 mm or greater, or reflecting telescope of 100 mm aperture or greater. A few organizations accept drawings with the disk diameter of 100 mm, which can be achieved with some binoculars.

In addition to the actual solar disk drawing, you should record the name of the observer, the equipment used, the observer's location (latitude and longitude), the date and time of observation (in Universal Time, UT), the weather and seeing conditions, and the relative sunspot number. For an example of an observing blank, see pages 48–49.

Daily geometry of the Sun

If you intend to draw the Sun, or to work out the position of any feature, you will need to take into account the Sun's changing aspect. If

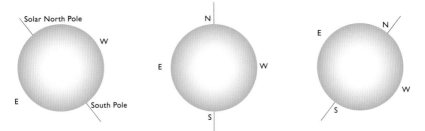

▲ The diagram shows how the Sun's inclination changes during the day, as seen from the northern hemisphere (although the movement of the solar axis has been exaggerated). Left is the Sun at sunrise; center is the Sun at midday; right is the Sun at sunset. Note that as you look at the Sun in the sky, the Sun's left-hand side is the east limb and the right-hand side is the west limb.

you look at the Sun in the sky from the northern hemisphere at about midday, the left-hand side is the eastern limb, while the right-hand side is the western limb. North is at the top and south at the bottom. Obviously, in the southern hemisphere everything is reversed: south is at the top, north at the bottom, the left-hand side is the western limb and the right-hand side the eastern limb.

The solar equator is tilted to the plane of the ecliptic (the plane in which the Earth orbits the Sun) by about 7.5°. In addition, the solar equator appears to tilt as the Sun moves through the sky during the day because of the Earth's tilt on its own axis. When the Sun sets or rises, the solar equator will be inclined to the horizon so the north pole of the Sun will not be at the top (or bottom if you are in the southern hemisphere) of the Sun as you look at it in the sky.

At sunrise from the northern hemisphere, the right-hand side of the solar equator will be higher than the left, so the north of the Sun is slightly left of the top of the Sun. As the Sun moves across the sky the equator becomes more horizontal, until at noon the Sun is overhead at the maximum altitude it will reach that day, with the solar equator horizontal and the north of the Sun at the top. As the day progresses, the Sun appears to tip the other way, with the right-hand side of the solar equator dropping down. At sunset in the northern hemisphere, the north of the Sun is slightly to the right of the top of the Sun as you look at it in the sky.

This daily change in geometry has an effect both on the paths sunspots take as they are carried round with the Sun's rotation, and on estimating the latitude and longitude of solar features. To make sense of any drawings, the orientation of the Sun at the time of observation should be noted. This is a fairly straightforward process which can be done at the time of observation (see below).

The aspect of the Sun also changes as the Earth moves in its orbit throughout the year due to the angle at which it is viewed. If you want to work out the position of any feature, or do separate sunspot counts for the northern and southern solar hemispheres, then the more complicated changing aspect of the Sun over the course of the year should also be taken into account. This can be determined from astronomical almanacs and is usually done after the observation is made. (See Chapter 5 for details of the yearly changing aspect of the Sun.)

Drawing the solar disk

Drawing the disk of the Sun is an excellent method of keeping a daily record of solar activity. You do not need to be a great artist, but as with all things, practice does help. By drawing the solar disk rather than just observing it, you will also improve your observation technique because you have to concentrate harder and for longer to make a drawing. You will start to notice features you did not notice before, and you need to process more visual information to reproduce an image than just to count spots.

The method of drawing will be determined by your observing equipment. Drawing directly on an observing blank with the Sun's image projected on to it is the easiest method. If you cannot draw directly on to your observing screen or on to an observing blank in a box, then you will have to use a graticule, a grid system that allows you to translate the positions and sizes of solar features from the projected image on to an observing blank.

I use a sharp pencil to draw sunspots, usually a harder one, such as H, for the outlines of the spots, and a softer, perhaps HB, to shade in the umbra and penumbra. I use an orange or yellow pencil to mark the positions of any faculae. To submit the drawings to an observing organization, you may want to take photocopies as it is impossible to replace drawings lost in the post. Pencil is not the best medium for photocopying, so you may want to try using a very fine black pen instead. Similarly, yellow pencil will not photocopy well, so I use parallel shading marks to distinguish the faculae. Experiment to find out which medium you prefer.

Drawing the solar disk directly

Drawing directly on to an observing blank is the easiest method of recording features on the Sun, but if you find it hard to begin with, do not despair! With a bit of practice you will find a method that suits both you and your equipment. Drawing methods can be very personal.

You first need to orient your observing blank or make a note of the inclination of the Sun's equator on your drawing to take account of the attitude of the Sun at the time of observing. There are various ways of doing this. One is first to draw a faint pencil line across the diameter of the solar disk from the designated east to west points on your observing blank. Then twist the sheet of paper until a feature such as a sunspot moves along the line with the motion of the Sun through the telescope. If you use a drive, you will have to turn it off to do this. Once a solar feature moves across the penciled diameter, the actual north pole of the Sun will be at 90° to the line, in line with the north pole of the Sun on the observing blank. The pencil line can be rubbed out afterward.

Then ensure that the image of the solar disk fits exactly into the circle of the observing blank. If the image fits exactly all the way round, it is projected on to a flat surface and has no distortion.

Keeping the image centered on the observing blank, quickly mark off the positions of the main features. If you do not have a drive on your telescope to follow the Sun's motion, just hold the telescope firmly and move it slowly to counteract the Sun's movement and keep the image in position. It is then possible to draw the features in detail. For complicated groups, you may need to hold the image over the observing blank to sketch out the main details.

If you have a complicated disk to draw, or spend some time drawing, check that the orientation of the disk is still correct. It does not take long for the Sun to move across the sky and as it does so, its inclination changes.

Drawing the solar disk using a graticule

If your equipment does not allow you to draw directly on to an observing blank, you will need to use a grid or graticule. Obviously the grid must be exactly the same size as the observing blank. An example of a grid is shown here, with a diameter of 75 mm. Photocopy it at 200% to use with a standard solar image of diameter 150 mm. Many

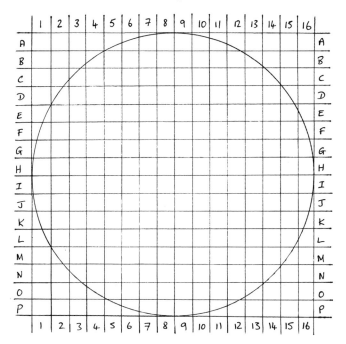

▲ In this example, the outline of the solar disk is partitioned by a grid. This allows any features to be copied accurately on to a separate piece of paper. If the grid is numbered and lettered at the top and bottom and both sides, it it easier to work out the correct box to copy the features into.

Solar rotation rates

The Sun rotates, but not as a solid body. It has what is termed differential rotation. This means that it rotates faster at the equator than near the poles. Features lying at different levels within the Sun also rotate at different rates, and different features can have different rotation rates even if they lie at the same level in the Sun and at the same latitude. Solar rotation is therefore not a straightforward phenomenon.

The fact that the Earth orbits the Sun complicates the issue further. It gives rise to two rotation rates for every feature: the synodic (with respect to the Earth) and the sidereal (with respect to the stars). Synodic rates are longer due to the Earth's orbit about the Sun.

If it were possible for an observer to be in a spacecraft a long way from the Sun, and not in orbit about the Sun, he or she would be able to record the sidereal rotation rate. If they were to observe a sunspot lying near the solar equator and time how long it took to make a complete revolution of the Sun, they would record a time of about 25 days. If an observer on the Earth watches the same sunspot and records how long it takes for the sunspot to return to the same place on the Sun as seen from the Earth, they will record a time of about 27 days.

The synodic rotation time as viewed from the Earth is longer because the perceived position of the sunspot on the Sun changes as the Earth moves along in its orbit. Whereas sunspots at the solar

▲ Imagine five sunspots lined up along the solar axis of rotation (above left). After one rotation (above right) of the Sun, the sunspot nearest the equator will have rotated farthest. Spots farther south or north will have rotated less far, depending on their distance from the solar equator.

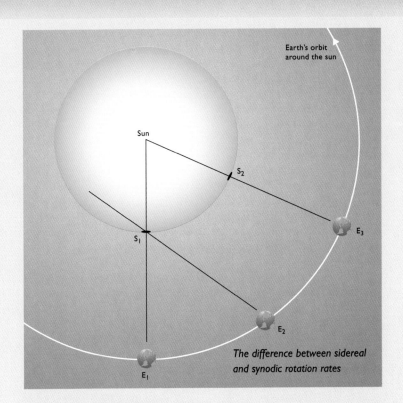

The difference between sidereal and synodic rotation rates

equator have a rotation rate of around 27 days (synodic) and 25 days (sidereal), any feature at or near the poles (for example polar faculae) will rotate in about 39 days (synodic) and 37 (sidereal).

The different types of sunspot rotate slightly differently: long-lived, regular shaped spots rotate slower than rapidly developing areas, for example. Rotation rates also vary according to the time in the solar cycle. When there are more sunspots and consequently more solar activity, features rotate slightly faster.

▲ *Imagine a sunspot at S_1, viewed from Earth at position E_1 in its orbit around the Sun. At this point, the sunspot appears in the middle of the Sun. It takes the sunspot about 25 days to reach position S_1 again. This is sidereal rate. When the spot has reached S_1 again, the Earth has moved in its orbit to position E_2. The sunspot does not appear to be in the middle of the Sun now because of the Earth's movement. The spot will not appear back in the middle of the Sun until the Earth has reached E_3 and the spot is at S_2. This rate is synodic and is about 27 days.*

observers design their own grid to fit in with their observing equipment and method.

All that is needed is a way of determining the position of features on the solar disk and a means of transferring them accurately to an observing blank. Thus whatever grid system is devised for the solar disk, it has to be replicated on the blank. One way of doing this is to draw the grid in very black ink which will show through a paper observing blank laid over it, allowing the solar features to be drawn in the correct position but not having any of the grid permanently on the observing blank.

The most difficult feature to transfer using this method of solar drawing is the inclination of the solar rotation axis to north, which depends on the Sun's position in the sky. With a static observing box it is generally not possible to twist the paper, as it is when drawing directly on to an observing blank. Some observers construct their observing box so that it can be twisted to allow a solar feature to run along one of the lines of latitude, thus determining the correct angle of inclination, but such a box is much more difficult to construct than one that does not move.

If you cannot twist your observing blank, you need to note the angle of inclination so that your final drawing can take it into account. One way of doing this is to allow a solar feature to drift across the grid, and then to make a faint pencil line from the center of the drawing blank at the angle at which the feature moves. This angle can then be translated into the correct inclination when the drawing is completed. To do this, first allow the solar image to move across the graticule until a feature lies at the junction of a line of latitude and a line of longitude. This gives a point of reference for the origin. The angle to the line of latitude at which the feature then continues to move across the observing blank is the angle you need to record.

After you have completed your drawing, the angle can be measured, and it is this angle that the solar north pole will make with the north pole of the drawing blank.

Once the inclination of the Sun in the sky has been determined, the features can then be drawn, using the grid references to reproduce their exact size and position. Again, if you have a complicated disk to draw, or if you need to spend some time drawing, check that the orientation of the Sun is still correct. It does not take long for the Sun to move across the sky, and as it does so its inclination changes.

Solar rotation

When Galileo and his contemporaries first used telescopes to observe the Sun, they saw sunspots that gradually moved across the visible solar disk with a period of roughly two weeks. After a great deal of discussion as to whether the sunspots were actually associated with the Sun or were instead something to do with the Earth's outer atmosphere, sunspots were used to estimate the solar rotation rate. The German astronomer Christoph Scheiner was the first to publish an estimate of the Sun's rotation rate. From the results of his observations of sunspots taken over the period 1611–27, he deduced that the Sun rotated in a period of 27 days.

The Sun is not a solid body. Sunspots are phenomena of the photosphere – sometimes called the solar surface – but they are not rigid objects lying on a solid surface. It is therefore not as easy to estimate the Sun's rotation rate from the motion of sunspots as it is by observing a surface feature on a terrestrial planet, for example. In fact solar rotation is a very complex subject (see the feature on pages 54–55), but if you track a sunspot from east to west across the Sun, it will take about two weeks, implying that the Sun takes about 28 days to rotate.

—— 5 · ANALYZING OBSERVATIONS——

If you are observing just for fun, you may not want to do any more than observe the Sun occasionally. However, because the solar disk changes on such a short timescale, it is easy to become fascinated by the daily (and yearly) changes. Once you start keeping a record of your observations, you may find you want to compare your observations, from day to day and also over the period of a solar cycle. There are various ways of interpreting your observations, and a simple way of making scientifically useful analyses.

If you just want to count the number of sunspots, then calculating the relative sunspot number is an easy way to be able to contribute to a professional database. Working out the relative sunspot number is easy, if sometimes time-consuming when there are more than a few hundred sunspots on the disk. To calculate the relative sunspot number, you do not even have to draw the solar disk.

If you want to draw the solar disk, thereby producing a more meaningful record of solar activity, all you need do is draw what you observe (see Chapter 4). However, if you want to compare your disk drawings either with previous observations, or drawings made by other observers, it will be necessary to account for the change in inclination of the Sun's axis as it moves through the sky during the course of a day. This is easy to do, as explained in Chapter 4.

In addition to the change in solar inclination during the day, the Sun's aspect changes throughout the year because of the changing angle at which an observer views the Sun as the Earth moves in its

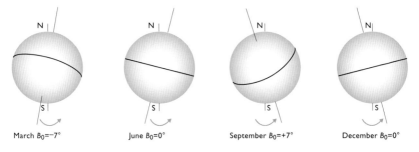

March $B_0 = -7°$ June $B_0 = 0°$ September $B_0 = +7°$ December $B_0 = 0°$

▲ The position of the equator changes through the year due to the Sun's "nodding" motion as Earth orbits it. In March the axis of the Sun is tipped the farthest away from the Earth so the lines of latitude curve downward. In June and December the Sun's axis is perpendicular to the Earth's orbital plane so the lines of latitude will be straight. In September the Sun's axis tips the farthest toward the Earth so the lines of latitude curve upward slightly.

orbit. The solar disk changes in diameter slightly, while the Sun's axis of rotation is seen at different angles, sometimes tipped toward the observer, sometimes tipped to the right, then away from the observer, then to the left. These changes affect the apparent positions of features as viewed from the Earth, but it is possible to correct observations to take account of the time of year.

Sunspots themselves change: appearing and disappearing, evolving and dying, producing an ever-changing spectacle. There are ways to classify sunspots, according to their size and shape, which indicate directly how active a particular sunspot group is, even though it is being observed only in white light. The area of the solar disk covered by sunspots is also a direct indicator of how active the Sun is.

Whatever type of solar observing you do, it is possible to make meaningful scientific analyses of your observations.

The nodding Sun

In addition to the change in the orientation of the Sun as it travels in its daily motion across the sky, the angle at which we view the Sun changes over the course of a year as we make one orbit on the Earth.

As we look at the Sun in the sky, we see it as a two-dimensional disk, so as viewed from our traveling vantage point the axis of the Sun appears to tip slightly from one side to the other. Over a period of six months, the axis of the Sun moves through 52.6°, from being inclined at 26.3° to the east to being inclined at 26.3° to the west.

In reality the solar axis does not just move from side to side, and the change in perspective as viewed from the Earth is three-dimensional.

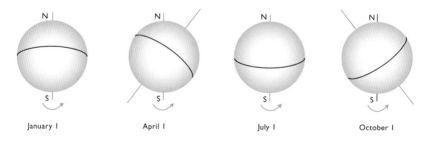

| January 1 | April 1 | July 1 | October 1 |

▲ The position of the equator changes throughout the year due to the Sun's sideways "tipping" motion as the Earth orbits it. This motion is combined with the Sun's "nodding" motion. In January and June the solar axis is perpendicular to the Earth's orbital plane so the lines of latitude appear to go from left to right across the Sun. In April the Sun's axis is tipped the farthest to the right, and in October it is tipped the farthest to the left.

Rather than simply moving from side to side as the Earth orbits the Sun, the Sun's axis traces out a circle. This means that sometimes the solar north pole is tipped toward the Earth, and sometimes away from it. This movement causes the solar equator to appear to move up and down.

The tilt of the solar axis is usually denoted by P, the position angle, and its day to day value is published by solar observing organizations. The position angle is the angle between the north point of the solar axis of rotation and the north point of the Sun as viewed in the sky. By convention, P is negative if the tilt is toward the western limb of the Sun, and positive if the tilt is toward the eastern limb.

The amount by which the solar rotation axis is tipped toward or away from the Earth is usually denoted by B_0, the heliographic latitude of the center of the disk, and again its value is published by solar observing organizations. B_0 varies from $-7.2°$ to $+7.2°$, a negative value indicating that the Sun is tipped away from us, causing lines of latitude to curve downward slightly and the heliographic latitude of the center of the disk to move up above the visual center of the disk. Conversely, a positive value indicates that the Sun is tipped toward us, causing lines of latitude to curve upward and the heliographic latitude of the center of the disk to move down.

All data such as the position angle, the heliographic latitude of the center of the disk and the heliographic longitude of the center of the disk are published by solar observing organizations and can also be found in annual publications such as *The Astronomical Almanac*.

The relative sunspot number

The relative sunspot number was introduced in 1848 by the Swiss astronomer Rudolf Wolf. It is still occasionally called the Wolf number. Wolf was investigating Heinrich Schwabe's discovery of the solar cycle and wanted a quick and easy way of noting how many sunspots were on the Sun and how they were grouped together. Although it is expressed by a simple empirical equation, the relative sunspot number has proved to be a very accurate way of indicating how active the Sun is, not only in white light but also across the whole electromagnetic spectrum.

Schwabe had also introduced the idea of sunspot groups. Sunspots do not appear randomly scattered over the solar disk, but generally appear in pairs or as members of a group. Some groups can become extremely complex, with many individual spots lying close together in the photosphere. Judging exactly what constitutes a sunspot group is an essential part of determining the relative sunspot number.

The number of sunspots has been recorded daily since 1845, and solar observing records have been examined before this date to extend the data backward. The relative sunspot number database is the longest running scientific database in existence.

The ease of calculating the relative sunspot number means that amateurs with inexpensive equipment can count the sunspots using this method and submit their results to observing organizations, which will then forward them to the professionals. It was something I found very satisfying when working professionally on data from the Yohkoh satellite (see Chapter 9). In the morning before work, I would use my small telescope and cardboard observing box to record solar activity and work out the relative sunspot number. I would submit my results monthly to the Solar Section of the British Astronomical Association, and at work I would use the amateurs' results.

Determining the number of groups

Where there is a group of sunspots, the magnetic field will be very intense – such areas are known as active regions. The word "group" for this purpose means anything from a single pore to an enormous complicated region containing hundreds of individual spots.

It is normally quite straightforward to decide how many groups there are: usually the regions are very distinct. On rare occasions it may be difficult to decide whether a scattering of spots constitutes one, two or more groups. The convention is to say that if spots lie within 10° of longitude of one another, then they belong to the same group. In general spots belonging to the same group will lie within a few degrees of the same line of latitude.

With practice, determining the number of groups can be done quickly by eye, as it is often obvious how many groups there are, with spots lying in distinct groupings, nicely spaced out from one another. However, it can be difficult to distinguish between groups when they lie near the solar limb: limb foreshortening means that 10° near the limb appears to be a shorter distance than 10° near the center of the disk. In addition, spots can lie close to the boundary of another spot or large group and it can be difficult to decide if they belong together.

Experience helps with counting groups, but the use of a template showing the effect of foreshortening will make the task very much easier. You will need a template of the same diameter as the solar image you are using. An example of one appears on page 64; it has a diameter of 75 mm, so photocopy it at 200% for use with a solar image of diameter 150 mm. You can also use Stoneyhurst disks to estimate distances of spots from one another, or Zürich grids (see pages 62–63).

As an example of estimating sunspot groups, consider a fictitious drawing of the Sun made on 13–13–01, shown on page 65. Let g be the total number of groups. For this drawing g is 16; of these, 9 lie in the

Stoneyhurst disks and Zürich grids

Stoneyhurst disks are very useful for determining the positions of features on the solar disk. They are also useful for determining whether sunspots are within the 10° limit that defines them as being part of the same group, no matter where they appear on the solar disk. They were originally produced by astronomers from the Stoneyhurst Observatory in England, and have subsequently been copied widely.

The complete set of Stoneyhurst disks consists of eight templates, each showing the latitude and longitude lines across the solar disk for different values of the heliographic latitude of the center of the disk (denoted by B_0) from 0° to 7° at intervals of 1°. In other words,

they show the latitude and longitude lines for the solar disk for $B_0 = 0°$, 1°, 2°, 3°, 4°, 5°, 6° and 7°. (The value of B_0 is published daily by solar observing organizations.)

Choose a disk with B_0 nearest the actual value for the time of observation. For example, if B_0 is +3.2°, select the Stoneyhurst disk corresponding to $B_0 = 3°$; if B_0 is 3.7°, the Stoneyhurst disk with $B_0 = 4°$ should be selected. It is a matter of choice if B_0 falls exactly between two Stoneyhurst disks, that is, if B_0 is 3.5°.

The same disk can be used for the same positive or negative value of B_0: the lines of latitude and longitude for negative B_0 are shown on the disk for the same positive value of B_0, and are read by turning the disk through 180°. For example, if B_0 is −3.2°, the Stoneyhurst disk with $B_0 = 3°$ would be selected, but it would be used upside down.

Zürich grids operate under the same principle but have much finer calibration. They are usually only calibrated between 50° north and south of the solar equator.

Both Stoneyhurst disks and Zürich grids

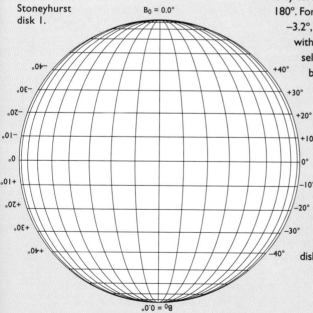

Stoneyhurst disk 1.

are very useful, especially if they are printed on transparent overlays which can be laid over a solar disk, enabling the grid reference of any solar feature to be read off straight away. The heliographic latitude and longitude are thus known immediately, and the Carrington longitude can be determined from the value worked out for the longitude of the central meridian, as explained later in this chapter. Alternatively, if the lines are produced in black enough ink, the disk or grid can be placed under the observation drawing and the lines of latitude and longitude viewed through the drawing.

The disks are reproduced here with a diameter of 75 mm. Photocopy them at 200% to produce disks for the standard diameter solar image of 150 mm.

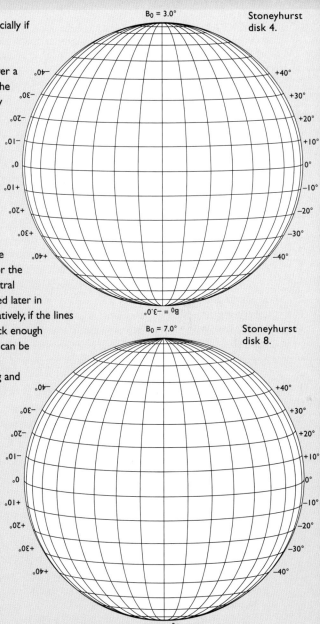

Stoneyhurst disk 4.

Stoneyhurst disk 8.

▲ *This template can be used for estimating limb foreshortening. Photocopy it at 200% to get a diameter of 150 mm.*

northern hemisphere, so $g_n = 9$, and 7 in the southern hemisphere, so $g_s = 7$. The groups have been numbered for ease of discussion.

The single pore, counted as group 1, appears to lie very close to the small pair labeled 2. In longitude they lie closer than 10°, even taking into account the fact that they are at the limb where the lines of longitude are foreshortened. However, they are counted as two distinct groups because of their difference in latitude.

The two spots in group 4 appear to be the same distance apart as the left-hand spot is away from the single spot designated group 3. However, the spot labeled 3 is considered a separate group because it lies on a different line of latitude from the two spots of group 4.

Take the number 8 Stoneyhurst disk, which shows the lines of latitude for a solar disk tipped away by seven degrees (B_0 for this fictitious day is −6.5°), and align the north and south points with the axis drawn in at a position angle of +10° in my drawing. You will easily see that spots belonging to group 3 lie on a higher line of latitude than do both spots in group 4, whereas both spots in group 4 lie very close in latitude. It is interesting to note how much the lines of latitude differ from the horizontal and what effect this has on the solar features, though of course this Stoneyhurst disk shows the maximum amount of tilt that can occur.

In a similar manner, groups 7 and 8 are separate because of their difference in latitude. In fact, these groups lie either side of the solar equator, group 7 lying in the northern hemisphere and group 8 in the southern hemisphere. This is reflected in the group counts for the different hemispheres (g_n and g_s).

The large single spots labeled 5 and 6 could be construed as a bipolar pair, except that they lie more than 10° apart in longitude. Even though there appear to be no other spots associated with either of these spots, they would not be counted as members of the same group.

In contrast, the group labeled 12, consisting of a pair of large spots similar to 5 and 6, do constitute a group because they lie closer than 10° in longitude. If you have a grid showing lines of longitude 10° apart, or if you have a Stoneyhurst disk or Zürich grid of the same diameter as the drawing of the solar disk, laying this over the drawing will show you that spots 5 and 6 lie more than 10° apart while the pair labeled 12 lie quite comfortably within the 10° limit.

If you position your grid at the correct angle, lining up the north point with the axis at position angle +10°, you will see that spot 5 and the left-hand spot of the pair labeled 12 both lie very close to the central

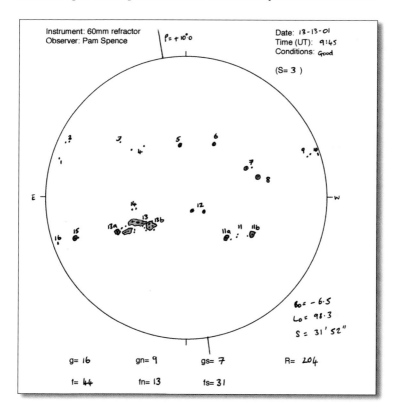

▲ A fictitious drawing made to represent various possible groupings of sunspots. In this drawing there are 16 groups, comprising 44 spots, so the relative sunspot number is 204. The drawing is shown here at a reduced size; photocopy it at 200% to produce a solar image of diameter 150 mm.

meridian, which is not apparent by examining the drawing without taking into account the tilt of the Sun's axis.

The two spots 5 and 6, and the two spots belonging to the pair labeled 7, lie close to the center of the solar disk, where the lines of longitude are at their greatest distance apart. It is interesting to compare them with the two spots labeled 15 and 16. Without the benefit of a grid showing the difference in distance between lines of longitude, a casual glance would indicate that spots 15 and 16 are a similar distance apart as the two spots in group 12 – after all, they appear a lot closer than groups 5 and 6. In fact, the spots labeled 15 and 16 are members of different groups because they lie more than 10° apart in longitude.

In a similar manner, at the western limb, the two spots of group 9 lie easily within 10° of longitude, but the spot labeled 10 lies more than 10° from the right-hand spot of the pair labeled 9. This may seem strange until you become familiar with the way the lines of longitude get progressively closer together at the limbs.

As you observe, you will become experienced in judging these distances, and if you had followed groups 9 and 10 across the disk during the previous two weeks, you would have seen they were separate groups. It is generally more difficult to ascertain group numbers at the eastern limb, where spots such as groups 15 and 16 appear, as you will not have been able to follow their progress across the reverse side of the Sun.

The very large group, 13, spans more than 10° in longitude, and spots 13a and 13b lie farther apart than 10°. However, as 13a and 13b belong to the same group, it does not matter that they lie more than 10° apart. In contrast, the two small spots making up the pair labeled 14 are counted as a separate group because they are definitely on a separate latitude from those in group 13.

All the spots within group 11 lie on the same latitude and so are counted as members of the same group, though it is less obvious that spots 11a and 11b belong to the same group than it is with spots 13a and 13b. Spots 11a and 11b would be taken as part of the same group because of the existence of the small spots between them and, again, because they lie on the same line of latitude. Once more, if you had observed this group as it moved across the solar disk, you would have seen the spots in the group closer together on previous days. If they had not been closer together on previous days, or had not had any small spots between them, they would have been counted as two separate groups, as with 5 and 6.

The relative sunspot number for this fictitious observation is 204, since the total number of groups is 16 and the total number of spots 44, giving $R = (10 \times 16) + 44 = 204$ (see below). This observation has been contrived to give examples of different spot and group configurations, but it is not impossible for the solar disk to appear like this. At solar maximum the disk can be a lot busier, and the relative sunspot number can be twice the value as in this example.

An empirical formula

The relative sunspot number (R) is given by the formula

$$R = 10g + f$$

where g is the number of groups and f is the total number of individual spots.

Thus, for example, if there are $g = 4$ sunspot groups and a total of $f = 23$ individual spots, the relative sunspot number, R, would be 63:

$$R = (10 \times 4) + 23 = 40 + 23 = 63$$

If only one spot is visible on the whole of the solar disk, the relative sunspot number would be 11, as the spot is first counted as a group in its own right, giving 1 multiplied by 10, to which 1 is then added for the total number of spots.

Despite being expressed by a simple empirical equation, the relative sunspot number does give an accurate indication of the Sun's activity, not only in white light but also in other radiation bands. This is because solar activity is very closely associated with sunspots and the total number of active regions on the solar disk.

If, for example, there are six sunspot pairs scattered over the solar disk, giving a sunspot number count of 72, the activity associated with them is roughly that associated with one complex group containing 62 spots (again giving a relative sunspot number of 72).

For more examples of determining R, see pages 68–69.

More sophistication

A few observers and observing organizations introduce a weighting factor into the relative sunspot number equation to take account of the means of observation and sometimes the observer. This factor is k (the reduction factor), and the relative sunspot number equation becomes

$$R = k(10g + f)$$

where f and g are defined as before. The reduction factor, k, is determined by how large an instrument is used for observation and how skilful the observer is. The larger the instrument and the more skilful the observer, the smaller k becomes. If you use a small instrument or are just starting to observe, k can be set at greater than 1.

In the past, some observers at the Zürich Observatory, where Wolf began his study of sunspot activity, have used a reduction factor of

Examples of working out the relative sunspot number

The relative sunspot number is given by $R = 10g + f$, where g is the total number of groups and f is the total number of spots.

Since the northern and southern solar hemispheres often have different activity cycles, with the maximum or minimum activity of a solar cycle occurring on different dates, observers often log the relative sunspot number for each hemisphere separately, denoting the number of groups and spots in the northern hemisphere as g_n and f_n, and in the southern hemisphere as g_s and f_s.

Example 1

An observation made on January 9, 1996, shows just one bipolar group fairly close to the western limb. There is one group ($g = 1$) and two spots ($f = 2$).

Therefore, the relative sunspot number for this day is $(10 \times 1) + 2 = 12$.

The group lies in the southern hemisphere, giving $g_s = 1$ and $f_s = 2$, so that the relative sunspot number in the south is $R_s = 12$. In this instance, of course, both g_n and f_n, and hence the relative sunspot number, R_n, for the northern hemisphere, are zero.

Note that the minimum value for the relative sunspot number is 0 (for a completely blank disk), but this jumps to 11 if there is just one spot because the individual spot is also counted as one group, so R would be $(10 \times 1) + 1 = 11$. The relative sunspot number will only lie between 0 and 11 if a reduction factor (k) of less than 1 is used.

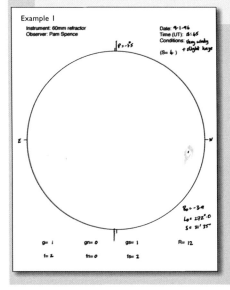

Example 1
Instrument: 60mm refractor
Observer: Pam Spence
Date: 9-1-96
Time (UT): 8:45
Conditions: very windy
(S= 6) + slight haze

$\rho = -55$

$\rho = -39$
$L = 272°.0$
$S = 31'35''$

g= 1   gn= 0   gs= 1   R= 12
f= 2   fn= 0   fs= 2

Example 3

An observation made on May 3, 2002, shows quite a busy disk. There are a few groups consisting of several spots, two bipolar groups and two individual pores. The three spots almost straddling the central meridian constitute a single group because they lie within 10° of one another (use a Stoneyhurst disk of the same diameter as the drawing of the solar disk to ascertain this to your own satisfaction). Although the two bipolar groups nearest the

Example 2

An observation made on May 18, 1995, shows two groups ($g = 2$), with several spots in both. In the group

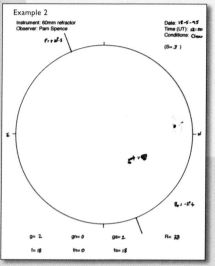

Example 2
Instrument: 60mm refractor
Observer: Pam Spence
P. r. of.5

Date: 18-5-95
Time (UT): 12:00
Conditions: Clear

($S= 3$)

B. : -2.4

g= 2   gn= 0   gs= 2   R= 38

f= 18   fn= 0   fs= 18

nearest the western limb there are 5 spots, and in the other group there are 13 spots, making a total spot count of $5 + 13 = 18$. Therefore, the relative sunspot number for this day is $(10 \times 2) + 13 = 38$.

Again, once the tilt of the Sun's axis of rotation is taken into account (position angle is $+20.3°$ and $B_0 = -2.4°$), it can be seen that both groups in fact lie in the southern hemisphere. Hence $g_n = 0$, $f_n = 0$ and $R_n = 0$; but $g_s = 2$, $f_s = 18$ and $R_s = 38$.

western limb appear the same distance apart, they do in fact constitute two groups because they lie more than 10° apart. This shows why limb foreshortening has to be taken into account when examining groups near the limb.

So, there are 10 groups ($g = 10$), and altogether there are 35 spots ($f = 35$). Therefore, the relative sunspot number for this day is $(10 \times 10) + 35 = 135$.

In the northern hemisphere there are 6 groups and 18 spots, giving $g_n = 6$, $f_n = 18$ and $R_n = 78$. In the

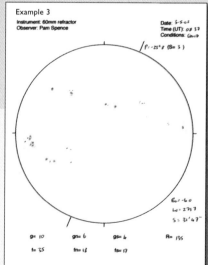

Example 3
Instrument: 60mm refractor
Observer: Pam Spence

Date: 3-5-02
Time (UT): 08 57
Conditions: Cloud

P: -25°8 (S= 3)

B. : -6.0
L. : 271.7
S : 31'47"

g= 10   gn= 6   gs= 4   R= 135

f= 35   fn= 18   fs= 17

southern hemisphere there are 4 groups and 17 spots, giving $g_s = 4$, $f_s = 17$ and $R_s = 57$.

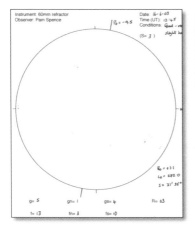

▲ *The drawing on the left shows my* *the same as those seen by the SOHO*
observation of the Sun made with a *spacecraft on the same day, as shown*
60 mm telescope. The major details are *on the right.*

significantly less than 1. Up until 1979, for example, values of k used at the observatory ranged between 0.58 and 0.63. One significance of this is that with k less than 1, the smallest relative sunspot number after zero can be less than 11.

In my own experience, however, the values of the relative sunspot number as determined by a wide variety of observers using a wide selection of instruments are normally very close – so close that I personally do not feel the necessity of using a weighting factor. This decision is reflected by many amateur organizations, which do not require the use of any weighting factor for any submitted results.

It is interesting to compare a drawing of the solar disk made with my 60 mm refractor and an image of the solar disk on the same day taken by the SOHO spacecraft. Apart from a few pores picked up by SOHO and not recorded in the 60 mm refractor, the images are very similar. The relative sunspot number for my drawing is 63 (five groups consisting of a total of 13 spots), while that for the SOHO image is 67 (five groups but 17 spots). Obviously the SOHO image does show much greater detail within the sunspot groups and also in the faculae. In general, the greater the magnification, the smaller the sunspots or pores that can be seen, but these small variations do not add greatly to the relative sunspot number, unless they constitute completely new groups.

The relative sunspot number can be calculated separately for each solar hemisphere, but to do this the position of the solar equator has to be determined by taking into account the change of perspective of the Sun through the year. This is explained in greater detail below, but it is still possible (by using Stoneyhurst disks) to determine each hemisphere's relative sunspot number without drawing the solar disk.

More numbers

Recording the relative sunspot number is a very useful method of seeing how the activity on the Sun varies from day to day, and professionals will use the daily number when investigating daily activity. However, for investigating activity over longer periods of time, for example over the 11 years of a solar cycle, it is more usual to use monthly mean daily frequencies (MDFs).

The mean daily frequency is data averaged over each month. For example, if you observed the Sun every day in July, to work out your mean daily frequency you would add up the relative sunspot numbers for each day and divide by 31 (the number of days in July).

If you only managed to observe the Sun for five days in December and had the results $R = 37, 49, 60, 55$ and 40, you would add up these values and divide by 5. Hence your MDF for December would be

$$\text{MDF} = \frac{37 + 49 + 60 + 55 + 40}{5} = \frac{241}{5} = 48.2$$

If you observed on a day when there were no sunspots, you would record a value of zero but would still count that day as one where you observed. Thus if you observed for three days in February and got the results $R = 21, 0$ and 12, your MDF would be

$$\text{MDF} = \frac{21 + 0 + 12}{3} = \frac{33}{3} = 11$$

As with the relative sunspot number, the MDFs for the different hemispheres can also be worked out.

Sunspot features

As with many astronomical features, sunspots have spawned their own terminology to describe the various structures observed within them. Professional (and some amateur) astronomers now observe in wavelengths other than white light, so this descriptive vocabulary has grown in recent years to accommodate features seen at other wavelengths, but the white-light features have retained their original names.

A "typical" sunspot pair consists of two dark umbral dots surrounded by a lighter, gray penumbra, but the morphology of sunspots and sunspot groups can be very diverse. The "front" spot of a pair, which leads the pair around the Sun with the Sun's rotation, is called

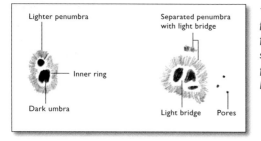

◄ The umbra and penumbra are the basic features of a large sunspot, but other features, such as light bridges, can also be seen.

the preceding spot (or p-spot), while the spot that trails is called the following spot (or f-spot). In general, the p- and f-spots will be of opposite magnetic polarity because they are the footprints of where the magnetic flux tubes arch in and out of the photosphere.

Small, single spots are termed pores. They can appear on their own, and also in pairs or groups. Pores do not possess a penumbra. If a penumbra appears they are termed spots rather than pores. Pores can be the first indication that a sunspot is to appear, but pores can also appear and disappear without evolving into sunspots. Pores are usually defined to be less than 10 arcseconds (10″) in diameter.

Void areas are often precursors to pores. They are areas in which the granulation is missing and which look slightly darker than the surrounding photosphere, though not as dark as pores. They do not last long, either disappearing or developing into pores. If you are to observe void areas, the seeing has to be very good.

"Sunspot" is the collective term used to describe a dark area on the Sun, but it is correctly defined as a dark umbral region greater than 10″ in extent, or any region showing an umbra and penumbra. A pair of sunspots is usually called a bipolar pair because each spot has a different magnetic polarity, though obviously observing in white light does not allow the observer to determine the magnetic polarities of sunspots. Regions where more than two spots appear within 10° of each other are called groups.

Umbrae are the dark inner regions of sunspots. They can vary in brightness: the darker the umbra, the stronger the magnetic field. For larger sunspots, a difference in brightness across the umbra can sometimes be observed, with a darker core surrounded by a slightly lighter outer region. Umbral dots can appear: tiny dots of umbrae slightly brighter than the surrounding umbrae. Umbral dots are regions of lower magnetic field within the umbra; they can have fields up to 1400 G weaker than their surroundings.

In contrast to umbral dots are so-called bright points, which also appear within umbrae. Unlike umbral dots, bright points are similar in structure to photospheric granulation and can form a network across the umbra. They are associated with the development of light bridges.

Light bridges are regions within both umbrae and penumbrae that effectively split a sunspot into two. The development of light bridges within a sunspot is closely related to the development of the sunspot itself. Most light bridges appear just after maximum activity within the sunspot and suggest that the spot is becoming unstable.

There are three main types of light bridge, though individual observers and observatories differ slightly in the way they classify them. Classic light bridges can last from a day to several weeks and consist of what appears to be normal photosphere seen through the umbra of a sunspot. Islands can last from a few hours to several days, and appear in the penumbra between the dark penumbra filaments. They do not appear to be linked directly to the photosphere. The third type, streamers, are often found within sunspot groups, existing within the umbra. They can last from several hours to several days.

Between the umbra and penumbra is a narrow zone known as the inner bright ring. This is a brightening of the penumbra, but it is different in origin from a light bridge. The outer bright ring lies on the outside of the penumbra, separating the penumbra from the surrounding photosphere. Material within the outer bright ring is at a higher temperature than the surrounding photosphere, up to 50 to 100 K hotter. The rings can cover large regions of the photosphere, up to two or three times the area of the whole sunspot itself. They are thought to be regions where the trapped energy beneath the sunspot is being diverted.

All sunspots, regardless of size, have outer bright rings, though they are not always easily observable. These rings have a distinct boundary with the surrounding photosphere, though the rings themselves can appear fragmented. An interesting observation is that faculae often surround the rings, but do not appear within them.

"Penumbra" is the term for the lighter, grayish surroundings of the umbra of larger sunspots. Even a small telescope will show detail within the penumbra, the most obvious being the filamentary structure radiating outward from the umbra. It was this structure that suggested to the US astronomer George Ellery Hale that sunspots were regions of intense magnetic field. The structure within the penumbra can change on a timescale of minutes, and is due to the flow of material along the magnetic field within the penumbra.

Sunspot sizes

In general, sunspots appear in pairs or in groups which can grow to be very complex, with many individual members. There appears to be a rough correlation between the area of the solar disk covered by sunspots and the activity of the Sun as indicated by the relative sunspot number. Thus any classification of sunspots by size will also give a classification by solar activity.

In general, the correlation between sunspot area and the relative sunspot number, R, is quoted as $16.7R$, which is to say that the area of the visible solar disk covered by sunspots is approximately 16.7 times the relative sunspot number (expressed in terms of millionths of the solar disk). It has been found that this relationship changes according to the stage of the solar cycle. At solar minimum, the ratio is generally less: the area covered by spots is approximately $11.5R$. The ratio is greater at solar maximum because there is a greater proportion of larger spots at this time.

Single pores have actual diameters of approximately 1000 km, though the most powerful telescopes can resolve single pores of around 300 km. The largest individual spots can have diameters up to 100,000 km, many times the diameter of the Earth, though it is rare for more than 1% of the visible solar hemisphere to be covered by spots at any one time.

It is usual to express sunspot areas in millionths of the solar disk. One millionth of the solar disk is equivalent to about 3,040,000 square kilometers (a spot with a diameter of about 1400 km). On the disk, that area would correspond to a small pore with an apparent radius of about 1.4 arcseconds (1.4″). Individual spots may grow to be as large as several thousand millionths, though typical small pairs have areas of around 100 or 200 millionths. A spot is visible to the naked eye if it covers an area of about 500 millionths of the solar disk.

The area of the solar disk covered by sunspots (the sunspot area number) is published daily by various professional solar observatories.

Estimating sunspot areas

The area of a sunspot is closely related to the magnetic field strength of the spot. This in turn is closely linked with any activity within the sunspot. Thus estimating the areas of sunspots in addition to the relative sunspot number provides another good indication of solar activity.

There are various methods of determining the area of a sunspot. Unfortunately it is not a straightforward operation as the apparent area covered by a spot depends on its position on the disk. Area estimates of spots near the limb have to have limb foreshortening taken into account. The same spot will appear to be much smaller near the limb than when it is straddling the central meridian. Limb foreshortening can have a very large effect on the apparent area of a spot, so if it is not taken into account, area estimates become meaningless.

Some organizations have produced templates that allow an almost immediate estimation to be made, but there is a fairly simple method of working out the area of sunspots, using a simple grid and a template to give the angular distance of the spot from the center of the Sun's disk.

First construct a grid of suitable size or use some appropriate graph paper. Use the same units as you have used for the diameter of your disk. For example, if you are using an observing blank with a radius of 75 mm, use a grid with 1 mm squares; the example shown here can be photo-copied at 200% to produce a grid for use with a image of radius 75 mm.

Your grid can be photocopied on to a transparency to be used as an overlay, or if the grid is marked out in dark enough ink, it can be used by laying the observation over the top of the grid. Count the number of squares each spot covers. Include any penumbra and count any squares which are more than 50% covered by the sunspot.

To calculate the angular distance of the spot from the center of the disk, another template can be used. Again, make your template the same diameter as your blank. The template is the same one as used for estimating the distance between groups. Estimate the angular distance of the center of the spot from the center of the disk.

Let r_0 be the radius of your solar drawing (in the same units as your grid, e.g. in millimeters).

Let A_M be the number of grid squares covered by the sunspot (this will be in the grid units squared, e.g. in square millimeters).

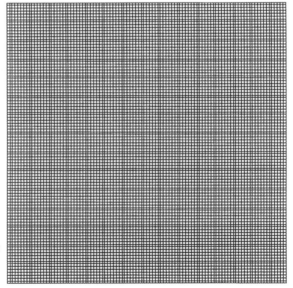

▶ A grid can be used to help estimate the area of the solar disk covered by sunspots. Ensure your grid is in the same units as your solar disk drawing.

Let δ be the angular distance on the surface of the Sun from the center of the disk to the group (in degrees).

Then A, the sunspot area in millionths of the Sun's visible hemisphere (corrected for foreshortening), is given by

$$A = \frac{A_M \times 10^6}{2\pi r_0 \cos\delta}$$

Example of determining sunspot area

As an example of determining the area of a sunspot and the total area of the solar disk covered by sunspots, see the accompanying fictitious observation. The drawing shows four sunspot groups. The solar disk

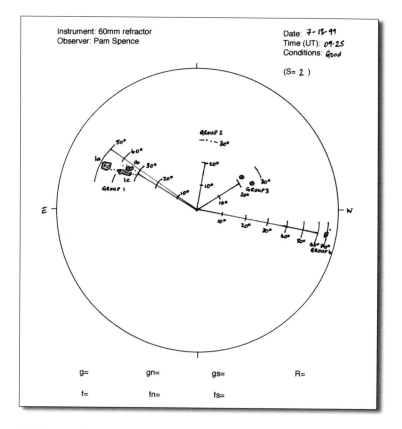

▲ A fictitious observation to represent examples of estimating sunspot areas. The original drawing was 150 mm in diameter. Grid lines have been drawn as examples, but a grid template would normally be used.

is 150 mm in diameter, making the radius 75 mm; a grid with 1 mm squares is used to estimate the sunspot areas. (The drawing reproduced here has a diameter of 75 mm; photocopy it at 200% to see it at its actual size.) Grid lines are shown radiating from the center and marked at 10° intervals. For an actual area estimation, the grid template would be placed under or over the drawing and the angular radial distance read off without recording it on the drawing.

The first step is to estimate the areas covered by the individual groups. **Group 1** consists of three large penumbral areas plus seven small spots. Each penumbral area should be estimated individually as the group itself spans more than 10° in longitude. Penumbral area 1a lies at an angular distance of 45° to 50°, while penumbral areas 1b and 1c both begin at an angular distance of 33°.

For penumbral area 1a, δ, the angular distance on the surface of the Sun from the center of the disk to the spot is about 45°. Holding the grid under the drawing shows that the spot covers 8 mm^2. Hence the quantity A_M for this spot is 8. Thus for this penumbral area, since the radius of the solar drawing, r_0, is 750 mm, the area in millionths of the Sun's visible hemisphere is

$$A = \frac{8 \times 10^6}{2\pi \times 750 \cos 45°} = 2400 \text{ millionths of the solar disk}$$

For penumbral area 1b, δ is 33° and A_M is 5, giving

$$A = \frac{5 \times 10^6}{2\pi \times 750 \cos 33°} = 1265 \text{ millionths}$$

For penumbral area 1c, δ is again 33° but this time A_M is 8, giving

$$A = \frac{8 \times 10^6}{2\pi \times 750 \cos 33°} = 2024 \text{ millionths}$$

The individual spots outside the penumbral areas can be estimated as covering about 1 mm^2 at a mean distance of 38°. If there were fewer tiny spots which together did not cover half a grid square, they would not be counted. In this instance, however, altogether they cover

$$A = \frac{1 \times 10^6}{2\pi \times 750 \cos 38°} = 269 \text{ millionths}$$

Thus the total area covered by this sunspot group is the sum of all the individual parts: 2400 + 1265 + 2024 + 269 = 5958 millionths of the solar disk.

Group 2 consists of two tiny pores lying at an angular distance of 30°. Together they would not cover half a grid square, so they are ignored.

Group 3 consists of two small spots which each cover about one grid square, one lying at an angular distance of 22°, the other at an angular distance of 25°. Each spot is estimated separately, giving areas of

$$A = \frac{1 \times 10^6}{2\pi \times 750 \cos 22°} = 229 \text{ and}$$

$$A = \frac{1 \times 10^6}{2\pi \times 750 \cos 25°} = 234$$

Thus the area of group 3 is 229 + 234 = 463 millionths of the solar disk.

It is interesting to note the difference in area caused by the difference in angular distance. Each spot of this group is estimated to cover 1 mm², and the only different quantity used in working out their areas is the very slight difference in angular distance: 22° compared to 25°. This 3° difference produces a difference of 5 millionths of the solar disk.

The fact that this formula is so very dependent on angular distance means that at the limb, where the lines of longitude are so close together, there can be large errors in determining sunspot area.

Group 4 lies close to the limb, and its angular distance is 70°. The penumbral area covers about 2 mm², giving an area of

$$A = \frac{2 \times 10^6}{2\pi \times 750 \cos 70°} = 1240 \text{ millionths}$$

Although penumbral area 1b is of a similar dimension (1265 compared to 1240 millionths of the solar disk), the apparent greater difference in area on the solar disk arises, of course, from the foreshortening of group 4 at the limb.

An amateur astronomer, Peter Meadows, has written a freeware program for PCs, called Helio, which calculates sunspot areas. The only inputs needed are the date and time, the position of the sunspot (*x* and *y* coordinates with respect to the center of the disk) and the number of grid squares covering the sunspot. The program is available for download from his website (see the bibliography).

Classification of sunspots

There are several schemes used for classifying sunspots. The most widely used are the Zürich or Waldmeier Classification Scheme (named for Max Waldmeier, who was Director of the Zürich Observatory between 1945 and 1979) and the McIntosh Classification Scheme (named for Patrick McIntosh of the US National Oceanic and Atmospheric (NOAA) Space Environment Laboratory), devised more recently during the 1990s. The McIntosh scheme is being used more and more, but if you wish to compare your observations to earlier ones, it may be helpful to use the Zürich scheme. See the features on pages 80–83 for details of the classification schemes.

Solar reference points

In order to discuss features on the Sun, astronomers need to know the solar latitude and longitude. This is not as easy as on a solid body like the Earth, as there is no constant reference point, and the latitude and longitude of solar features change as the Sun rotates and as the Earth moves in its orbit. However, it is possible to allocate latitude and longitude to solar features, based on a system devised in the 1850s by the British astronomer Richard Carrington. This system deals with Carrington rotations, which are explained below.

Peter Meadows' Helio program, mentioned above, can be used to determine the heliographic latitude and longitude of sunspots. The only inputs needed are the date and time, and the position of the sunspot (x and y coordinates with respect to the center of the disk).

Solar latitude

Latitude is slightly easier to determine than longitude. As on Earth, latitude is determined north or south of the equator, being 0° at the solar equator and 90° north or south at the poles. Once you have determined where the solar equator is, it is fairly easy to work out the latitude of a solar feature such as a sunspot. Limb foreshortening causes the lines of latitude to appear closer together at the poles, just as the lines of longitude appear closer together toward the limbs of the Sun, but because sunspots rarely appear more than 40° from the solar equator, this does not generally affect the determination of the solar latitudes of sunspots.

The difficulty arises in determining the position of the solar equator. Not only do you have to eliminate the effect of the Sun's inclination in the sky due to its daily movement, but you also have to know the position of the Sun's axis of rotation. However, these difficulties are easily surmounted.

Taking account of the angle of inclination of the solar equator because of the Sun's motion across the sky is straightforward. Imagine a sunspot on the solar disk. No matter at what latitude it lies, by definition, the line of latitude it does lie on will be parallel to the solar equator.

When you have projected the Sun's image into your observing box or on to your screen, the Sun's natural motion across the sky will move it fairly rapidly across the card or paper. If you have a telescope with a drive, you will have to turn it off to see this effect. It is quite amazing how fast the Sun's image will move – annoying when you are trying to count or draw features, but this movement makes it easy to determine the angle at which the equator and lines of latitude are tilted from the horizontal.

If you let your solar image drift across the paper, it will naturally drift from east to west because of the Earth's rotation. The sunspot will therefore follow a line across the paper which is parallel to the solar equator. Thus observing the line that the spot follows will give you the inclination of the line of latitude and hence the inclination of the solar equator.

With my observing system, I draw a faint pencil line from east to west across the diameter of my observing blank. I then pin the paper into my cardboard observing box and move it until a sunspot or other solar feature moves along the line. I then secure the paper with another pin and make my observation or drawing, knowing that the observing blank is now tilted at the correct angle of inclination. Once the drawing is finished, I rub out the pencil line.

With the published values of the position angle, P (the inclination of the Sun's axis of rotation to north), and B_0 (the amount the Sun's axis

The Zürich (or Waldmeier) Sunspot Group Classification

First published in 1938, the Zürich or Waldmeier classification scheme for sunspots had eight categories. One further category was added in 1939. This scheme classifies a sunspot group by a letter for the class of group and the number of individual spots within the group.

The nine classes of group are:

A An individual spot or a group of spots without a penumbra or bipolar structure.

B A group of sunspots without a penumbra or bipolar structure.

C A bipolar sunspot group, the principal spot of which is surrounded by penumbra.

D A bipolar group, the principal spots of which are surrounded by penumbrae. At least one of the two principal spots has a simple structure. The group is in general less than 10° across.

E A large bipolar group in which the two principal spots are surrounded by penumbrae and generally exhibit a complex structure. Numerous smaller spots lie between the two principal spots. The group is at least 10° across.

F A very large bipolar or complex sunspot group with a length of at least 15°.

G A large bipolar group without any small sunspots lying between the two principal spots and a length of at least 15°.

H A single spot with penumbra and diameter of greater than 2.5°.

I A single spot with penumbra and diameter of less than 2.5°.

of rotation is tilted to or away from the Earth), I can work out the position of the solar equator and hence the latitude of any solar feature.

I also make use of a set of Stoneyhurst disks (see the feature on pages 62–63), which have predrawn lines of latitude and longitude for particular values of the tilt of the Sun's axis; they can be laid over a drawing to give the position of the solar equator and lines of latitude.

Solar longitude

Not having any static reference point on the Sun makes determining solar longitude difficult. As with any roughly spherical body, there is not even a preferential line that can be taken as zero longitude, as there is with the equator for latitude. Unfortunately, there are also no small land masses with astronomical observatories situated at places called Greenwich, which would allow us to define zero longitude on the Sun.

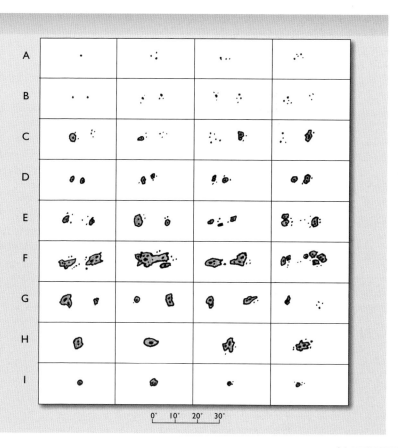

As we look at the Sun in the sky or at a projected image, the line that can be drawn through the north and south points, crossing the center of the disk, is called the central meridian. The longitude of the immediate solar disk can be described as degrees east and west of this line, but this does not help define the position of a particular feature over time or from observer to observer, because the solar disk is constantly moving.

For example, if I observe a sunspot 10° west of the central meridian at 09:40 hours on June 24, 2003, anyone else observing at this time and

The McIntosh Sunspot Group Classification

The McIntosh sunspot classification scheme modifies and extends the Waldmeier classification. Sunspot groups are classified by a three-letter code. The first letter describes the group type (based on a modified version of the Waldmeier scheme). In the McIntosh scheme, the Waldmeier groups G and I are dropped; group G is now included in group E or F, while group I is included in group H. The second letter describes the penumbra of the largest spot of the group, while the third letter describes the compactness of the spots in the immediate part of a group.

Group type

A An individual spot or a group of spots without a penumbra or bipolar structure.

B A group of sunspots without a penumbra or bipolar structure.

C A bipolar sunspot group, the principal spot of which is surrounded by penumbra.

D A bipolar group, the principal spots of which are surrounded by penumbrae. At least one of the two principal spots has a simple

structure. The group is in general less than 10° across.

E A large bipolar group in which the two principal spots are surrounded by penumbrae and generally exhibit a complex structure. Numerous smaller spots lie between the two principal spots. The length of the group is at least 10° across.

F A bipolar group with penumbrae on spots at both ends of the group and more than 15° long.

H A unipolar group with penumbra.

Penumbra of largest spot

x No penumbra (class A or B).

r Rudimentary (incomplete) penumbra partly surrounding largest spot; irregular boundaries with width only around 0.2°. Penumbrae brighter than usual with granular fine structure.

s Small, symmetric penumbra, elliptical or circular, with filamentary structure and less than 2.5° long.

a Small, asymmetric penumbra, irregular in outline, with filamentary structure and less than 2.5° long.

date should see it in the same position. However, if I, or anyone else, observes again a few days later, the Sun will have been rotating during that time, and the sunspot will have moved and will no longer be 10° west of the central meridian. It would thus be meaningless for me to try to talk to another observer about a feature seen 10° west of the central meridian unless they had also observed at that same time.

To overcome this problem, the English astronomer Richard Carrington arbitrarily defined a zero longitude as the meridian crossing

h Large, symmetric penumbra, more than 2.5° long.
k Large, asymmetric penumbra, more than 2.5° long.

Spot compactness

x Individual spot.
o Open distribution with few if any spots between principal spots; the group clearly consists of two parts.

i Intermediate distribution with numerous spots between principal spots, all without mature penumbra.
c Compact distribution with many large spots between principal spots, with at least one with a penumbra.

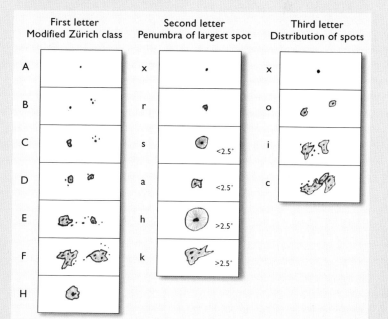

the center of the visible disk on November 9, 1853. Since then this meridian has been carried around with the solar rotation. Rotations are counted from that time, with the longitude of the central meridian decreasing from 360° to 0° degrees during each rotation as the central meridian point rotates with respect to the Earth. Carrington rotation rates are defined as 25.38 days (sidereal) or 27.2753 days (synodic), and from these values the longitude of the meridian passing through the center of the disk at any time can be referred to the position of zero longitude. More recently, this zero meridian was redefined as the meridian that passed through the ascending node of the Sun's equator on the ecliptic at 12:00 UT on January 1, 1854.

L_0 is the heliographic longitude of the center of the disk. Like the position angle and the heliographic latitude of the center of the disk, it is published daily by solar observing organizations. L_0 decreases by about 13.2° per day.

Just as with solar latitude, Stoneyhurst disks (see pages 62–63) can be used to read off the longitude of any feature east or west of the central meridian. Then by referring to L_0, the actual Carrington longitude of the feature can be worked out.

Mathematical method of determining latitude and longitude

From a solar drawing, it is possible to determine the heliographic latitude and longitude of a solar feature such as a sunspot directly by mathematical means.

Imagine that there is a sunspot at position x, at angle θ from the uncorrected position of the north point of the solar disk – the north point of the solar observing blank. The angle θ is measured from north, through east, south and west back to north, through to 360°. So if the spot is in the western hemisphere, for example, the angle will be greater than 180°. Note that my observing blank shows the solar disk as it appears in the sky with west to the right, because I use a star diagonal, so the angle θ on my disk will be measured counterclockwise.

Let the radius of the solar disk be r_0 (usually around 75 mm) and let the distance from the center of the disk to the sunspot at x be r (the radial distance). Note that the distance r must be measured in the same units as the radius of the disk (i.e. if the radius is 75 mm, r must also be in millimeters), otherwise the ratio r/r_0 will not be correct.

The Sun's angular diameter changes over the course of the year, and its value, S, for the day of observation can be found from the same published source as the position angle, P, the heliographic latitude of the center of the disk, B_0, and the heliographic longitude of the center of the disk, L_0.

The actual position angle of the spot on the solar disk, α, is given by

$$\alpha = \arcsin(r/r_0) - S(r/r_0)$$

Note that if the angular radius (R_S) of the Sun is used instead of the diameter, it is necessary to replace S by $2R_S$ in the above equation.

The heliographic latitude, B, of the sunspot is then found from

$$\sin B = \cos \alpha \sin B_0 + \sin \alpha \cos B_0 \cos(P - \theta)$$

and the difference between the sunspot and the central meridian, L_D, from

$$\sin L_D = \sin(P - \theta) \sin \alpha/\cos B$$

Whether L_D is positive or negative determines its position relative to the central meridian. If the spot is west of the central meridian, L_D is positive, and conversely if the spot is east of the central meridian, L_D is negative.

The Carrington longitude, L, of the sunspot is then found from

$$L = L_0 + L_D$$

Be careful of the sign of L_D when determining the Carrington longitude.

Example

Consider the fictitious observation shown on page 86 as an example. The solar image is projected into a 150 mm diameter circle (note that the illustration is reproduced here at a reduced size), so the radius of the solar disk, r_0, is 75 mm. From a standard source, the position angle on this day is found to be +15°, the heliographic latitude of the center of the disk, B_0, is +2.3°, and the heliographic longitude of the center of the disk, L_0, is 139.3°. Since my observation was made through a star diagonal, the angles to the groups are measured counterclockwise.

The angular diameter of the Sun on our fictitious day is 31′ 39″. To use this value, it is easiest first to convert it into degrees. 39″ is 39′/60 = 0.65′. Thus 31′ 39″ = 31.65′. This in turn is 31.65°/60 = 0.5275°.

Consider spot group 1. All measurements are made to the middle of the group. Thus the angle θ from the corrected position of the north point of the solar disk to the group is 45°, and the radial distance from the center of the disk to the center of the group is measured to be 45 mm.

Thus the actual position angle of the group on the solar disk, α, is given by

$$\alpha = \arcsin(r/r_0) - S(r/r_0)$$

so $\alpha = \arcsin(45/75) - 0.5275(45/75)$

$$= 36.87 - 0.3165 = 36.55$$

The heliographic latitude, B, of the sunspot is then found from

$$\sin B = \cos \alpha \sin B_0 + \sin \alpha \cos B_0 \cos(P - \theta)$$

so $\sin B = \cos 36.55° \sin 2.3° + \sin 36.55° \cos 2.3° \cos(15 - 60)°$

so $\sin B = 0.03224 + 0.42076 = 0.45300$

giving $\sin B = 0.4530$ and $B = 26.94°$

Then the difference between the sunspot and the central meridian, L_D, is given by

$$\sin L_D = \sin(P - \theta) \sin \alpha / \cos B$$

so $\sin L_D = \sin(15 - 60)° \sin 36.55° / \cos 26.94°$

giving $\sin L_D = \sin(15 - 60)° \sin 36.55° / \cos 26.94° = -0.4724$

and $L_D = -28.2$

The Carrington longitude, L, of the sunspot is then found from

$$L = L_0 + L_D$$

so $L = 139.3 - 28.2 = 111.1$

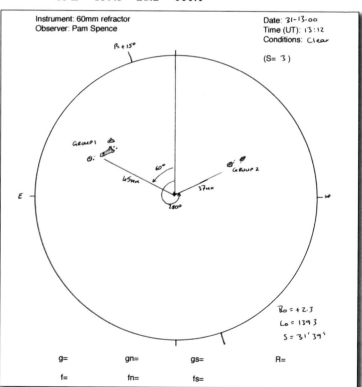

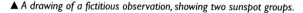

▲ A drawing of a fictitious observation, showing two sunspot groups.

6 · SOLAR ECLIPSES

A total solar eclipse is an awe-inspiring experience, and if you ever have the chance to observe one then grab it with both hands. There are three types of solar eclipse: a partial eclipse, when the Moon covers only part of the Sun; an annular eclipse, when the entire Moon is silhouetted against the disk of the Sun but does not cover all of the Sun's photosphere; and a total eclipse, when the Moon blocks out the entire photosphere. A partial or annular solar eclipse is interesting, but the difference between these and a total eclipse is like the difference between a village pond and Niagara Falls!

When and why eclipses occur

As the Earth orbits the Sun, the Moon orbits the Earth, and occasionally Sun, Earth and Moon will lie in the same plane and an eclipse will

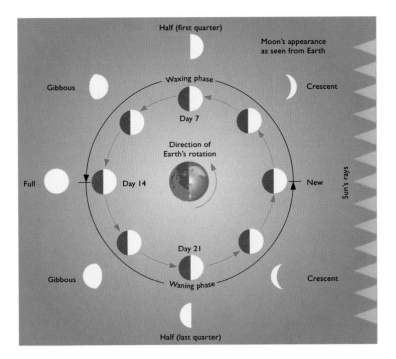

▲ Everyone is familiar with the changing phases of the Moon. They are caused by the Moon orbiting the Earth and being seen at different angles with respect to the Sun. The Moon can only be seen by reflected light from the Sun so at different times, we see different portions of the Moon illuminated.

take place. When the three bodies are all in the same plane and the Earth lies between the Sun and the Moon (at the full moon stage of the monthly lunar cycle), the Moon passes through the shadow cast by the Earth and a lunar eclipse occurs. When the Moon passes between the Sun and the Earth, the Moon cannot be seen as all the Sun's light falls on the side of the Moon facing away from Earth. This stage in the monthly lunar cycle is called new moon. If at this time the Moon's shadow falls upon the Earth, a solar eclipse occurs.

A tilted orbit

The Moon takes about a month to orbit the Earth, giving us the phases of the Moon with which we are all familiar, but eclipses do not happen every month because the Moon's orbit is tilted by about 5° (to be precise, 5° 8′ 43″) to the ecliptic (the plane of the Earth's orbit about the Sun). This means that the Moon sometimes lies above or below the Sun in the sky at new moon, and above or below the Earth's shadow at full moon, and an eclipse does not occur.

This slight tilt of the Moon's path about the Earth means that it passes through the ecliptic plane twice during each orbit about the Earth. The points where the Moon crosses the plane of the Earth's orbit are called nodes. When it crosses from south to north, the point it passes through is called the ascending node; when it crosses from north to south, it passes through the descending node.

The positions of these nodes slowly move around the Earth because of the gravitational influence of the Sun and the Earth on the Moon. This movement is called the regression of the nodes. Only when a node is close to new or full Moon can eclipses take place. The line of nodes – the imaginary line joining the ascending and descending nodes – completes a revolution in 6585.32 days (about 18 years 11 days, depending on how many leap years are included). This period is known as the saros.

A remarkable coincidence

It is one of the most amazing coincidences that the diameter of the Sun and the Moon appear roughly the same in the sky as viewed from the Earth. The actual diameter of the Sun is around 400 times that of the Moon (1,391,000 km as compared with 3476 km), but because the Sun happens to lie about 400 times farther away than the Moon,

APPARENT DIAMETERS OF THE SUN AND THE MOON			
	Maximum	Mean	Minimum
Sun	32′ 35″	32′ 01″	31′ 31″
Moon	33′ 31″	31′ 05″	29′ 22″

their apparent diameters in the sky are very similar. This means that when the Moon lies directly between the Earth and Sun, it is sometimes the correct size to block out the light from the photosphere, allowing the tenuous white outer atmosphere of the Sun, the corona, to be seen.

A changing scenario

As well as the Moon's orbit about the Earth being tilted to the plane of the Earth's orbit about the Sun, the Moon's orbit about the Earth, and the Earth's orbit about the Sun, are not completely circular. So even if an eclipse does take place, occasionally the Moon's disk in the sky will not be quite large enough to block out the photosphere, and the corona will not be seen. These eclipses are called annular eclipses, and a ring of photosphere – sometimes called the ring of fire – will surround the dark disk of the Moon.

The period of annularity can last as long as 12.5 minutes, but usually it is much shorter. An eclipse can start off being annular, then become total when the umbral shadow (the shadow cast when the

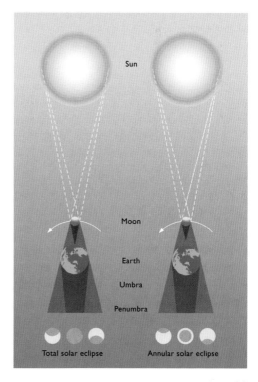

▶ In both these solar eclipse scenarios the Sun, Earth and Moon all lie in the same plane. In the left-hand diagram the Moon's shadow reaches the Earth's surface so a total eclipse is seen. In the right-hand diagram the Moon's shadow falls short of the Earth's surface so an annular eclipse is seen.

Sun

Moon

Earth

Umbra

Penumbra

Total solar eclipse   Annular solar eclipse

Moon completely blocks out the Sun's photosphere) reaches the Earth's surface. When the Moon is only just too small to cover the Sun's disk, the eclipse is called an annular-total.

Again, due to the different distances of the bodies, the totality of a total eclipse can last different times depending on how large the Moon's disk is compared with the Sun's. The larger the relative size of the Moon's disk, the longer totality lasts. The longest it can ever last is 7 minutes 31 seconds, but it is usually much shorter than this.

The frequency of eclipses

The geometry of the orbits of the Sun, Earth and Moon produces up to seven eclipses (both lunar and solar) in any one year, but only a maximum of five solar eclipses can happen. It is rare to have a year with seven. A table of forthcoming eclipses can be found at the back of this book.

Since eclipses only happen when either full or new moon occurs at one of the nodes, they often take place in pairs. The line of nodes intersects the Moon's orbit at points 180° apart. So if one node is near new Moon, the other node at that time will be near full Moon, 180° away. So if a lunar eclipse occurs (when the Moon is full), a solar eclipse will usually precede or follow in about 15 days (when the Moon is new).

Shadow on the Earth

The shadow cast by the Moon on the Earth has two components. The dark central zone is called the umbra, where none of the light from the photosphere reaches the Earth, and the outer region is the penumbra, in which a portion of photospheric light reaches the ground.

Only observers in the umbral shadow will see a total eclipse. Observers in the penumbral shadow will see a partial eclipse. Thus even if an eclipse is total, not everyone will see a total eclipse – it will depend where the observer is located. The umbral shadow is quite small compared to the surface of the Earth, so only a few lucky people will see the total phase. More will see a partial phase, but many will see no type of eclipse at all. This is in contrast to lunar eclipses, which are visible from any point on the Earth from where the Moon is visible in the sky.

The umbral shadow will move in a west-to-east path across the Earth's surface. This narrow curved zone, called the path of totality, can be up to 269 km wide, but is usually a lot less, averaging about 160 km. The path of the Moon's shadow travels in curved trajectories several thousand kilometers long over the surface of the Earth, but because of the nature of the Earth's surface, the path of totality can cross inhospitable land and vast tracts of sea, which adds to the rarity of the possibility of observing these events.

The probability of a total solar eclipse occurring in the same place on Earth is about 1 in every 370 years. However, many people who have had the opportunity to view a total solar eclipse will make every effort to travel to see another one – such is the impact of the phenomenon.

Equipment for eclipse observing

The only absolute prerequisite for eclipse observing is a safe way to watch the partial or annular phase. If you are preparing to watch a total eclipse, station yourself as near the central line (the middle of the umbral shadow) as possible at a site where you can see the Sun at the appropriate time. This may sound obvious, but eclipses can happen at any time of the day, so if it is to take place near sunrise or sunset you will need a clear, low horizon in the appropriate direction. If you are traveling to unknown territory, ensure that no nearby trees or buildings, or distant hills or mountains, will block your view at the critical moment.

Once your observing site is chosen, make sure you will be comfortable. Set up any equipment you have for observing the Sun by projection, or for photographing the Sun directly, well ahead of the eclipse. If you are in a hot climate, have plenty of water and protection for yourself and your equipment from the heat. Conversely, if you have traveled to a cold site, ensure that you have enough clothing and warm drinks for yourself and protection for your equipment from the cold. Eclipse-watchers will tell you that if anything can go wrong, it will – so prepare as much as possible for all eventualities.

You may like to take a small tape recorder to record your own reactions and those of anyone nearby. This may seem strange, but I can assure you that the onset of totality will cause people to react in very unusual and verbal ways! If you are attempting to photograph the eclipse, the tape recorder can help you to record what you are doing, without you having to stop and write notes.

A white sheet can help you see the partial phase, shadows of the partially eclipsed Sun and shadow bands (see below). Put it on the ground under a tree, for example, where the leaves act as mini pinhole cameras and the shadows appear as partial solar disks. An accurate watch will help you be prepared for each event during the eclipse, and if you are taking any sort of camera, make sure you have enough film and batteries. Totality is not the time to run out of either of these!

Eclipse filters

As with observing the everyday Sun, great care must be taken when observing an eclipse. Until the photosphere is covered, the Sun must be treated as the dangerous object it is and must never be looked at directly unless adequate eye protection is used. The only time it is safe to look directly at the Sun is at totality, when the harmful photospheric light is blocked by the Moon and the solar corona is on view. This means that during a partial or an annular eclipse, adequate eye protection must be worn at all times if the eclipse is viewed directly.

The partial phase of a total eclipse, a partial eclipse or an annular eclipse can be viewed safely by projection. Chapter 3 describes how to do this with binoculars or a telescope, while Chapter 2 describes how to make a simple pinhole camera, which will allow the partial phase to be observed safely.

If you observe the Sun directly at other times, then your filter equipment should be adequate to observe the partial and annular phases safely. Check there is no damage to your filter before using it, particularly if you have had to transport it any distance.

Many cheap filters and "eclipse glasses" will be on sale just before an eclipse. My advice is not to use them. Even though some will be made from genuine solar filter material such as Mylar, some will not. Unless you are able to verify that the filter you have bought is 100% safe, do not use it. It is not worth running the risk of damaging your eyes.

A total eclipse

If you are lucky enough to be able to experience a total solar eclipse, you may want to spend the time just observing this amazing phenomenon, particularly if totality is short. There is nothing sadder than someone trying to take a photograph of totality and spending all the time behind a camera lens and missing out on the myriad of events that surround totality.

The partial phases, however, can take up to two hours to complete, so there is plenty of time to take a photograph. The only problem with the partial phase is that the light from the photosphere is still visible, so it is not safe to look directly at it without suitable filters. Be very careful and never look directly at the Sun until all of the photosphere is fully covered by the Moon. Use the methods suggested for viewing or photographing the everyday Sun, as described in Chapter 7.

A total eclipse has four stages, defined by the positions of the Moon and Sun.

First contact

The moment when the Moon takes the first tiny bite out of the Sun's disk is called first contact. This corresponds to the time when the Moon's penumbral shadow first covers the observer. Since the width of the penumbral shadow is greater than that of the umbral shadow, the partial phase of a total solar eclipse can be seen over a much

The stages of a total solar eclipse

▼ **First contact** occurs when the Moon's edge (limb) first starts to cover the solar disk.

▼ **Second contact** is just about to occur, when the Moon's eastern limb reaches the Sun's eastern limb. Totality begins.

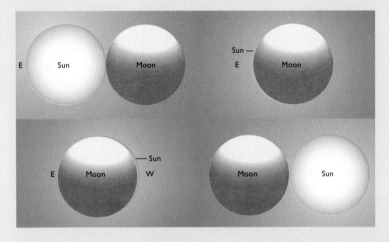

▲ **Third contact** has just occurred, when the Moon's western limb leaves the Sun's western limb and the Sun's photosphere reappears at this side. Totality is over.

▲ **Fourth contact** occurs when the last of the Moon finally leaves the solar disk. This marks the end of the solar eclipse.

wider area than the total phase. For example, at the total solar eclipse visible from the southernmost tip of England on August 11, 1999, people in the north of Scotland, nearly 1300 km away, could see the partial phase.

Observers not in the path of totality will see only a partial eclipse. How far they are from the path of totality will dictate how great or small a partial eclipse they will see. The magnitude of eclipse is the proportion of the Sun's disk that is covered by the Moon. Again, taking the eclipse that was visible from the UK in 1999 as an example, people in Inverness in the north of Scotland could see an 80% partial eclipse: in other words, 80% of the photosphere of the Sun was covered by the Moon. For observers in London, 924 km farther south, the Moon covered 96.6% of the Sun's disk. Despite being so close to 100%, this was still a partial eclipse because it only needs a tiny part of the photosphere's light to be visible to block out the ephemeral light of the corona, which is the awesome trademark of a total solar eclipse.

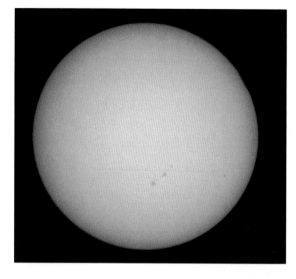

◄ *In this image of first contact, the lunar disk is just visible at about 2 o'clock on the Sun's disk. At this point a tiny indent in the Sun's outer limb can be seen.*

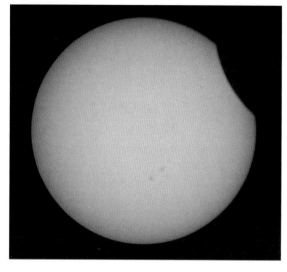

◄ *The partial phase continues as the Moon slowly covers the solar disk.*

The partial phase

The Moon appears to move very slowly across the Sun until the crescent Sun becomes very thin. From this point everything seems to happen very quickly, but there is plenty to watch during the apparently long partial phase. On average, the first partial phase will last slightly under an hour.

As the partial phase deepens, take a look at shadows cast by leaves on a tree, or by anything else that has a very small gap through which the sunlight falls. The dots of sunlight will show the partial phase as the gaps between the leaves are acting as tiny pinhole cameras. These partial Sun-shadows are easily photographed – just make sure that your camera flash is turned off or disabled.

If there are any sunspots on the disk of the Sun during the eclipse, it is fun to watch the Moon slowly gobble them up. "Serious" science can also be done by timing the Moon's passage across any large active areas. These timings, along with a record of where on the Sun these regions lie and a knowledge of the distances of the Sun and Moon from the Earth, will give an approximate size for the sunspots.

As the partial phase grows, animals and birds – taken rather unawares – suddenly decide that night is coming. Birds stop singing and animals prepare for bed. At one eclipse I was lucky enough to observe in India, vultures suddenly flew back to roost in a nearby tree. Nocturnal animals begin to stir, but the event is usually over too fast for them actually to appear.

Baily's beads and the diamond ring

The Moon is not smooth, like a billiard ball. Its surface is covered in valleys and mountains, which can appear at the edge of the Moon's disk as we look at it in the sky. As the Moon's disk crosses the Sun, these valleys and mountains at the lunar limb will appear silhouetted (through suitably filtered equipment). If you observe the partial phase directly with a magnification of about ×100 or higher, you should be able to see the Moon's profile in detail. Data published at

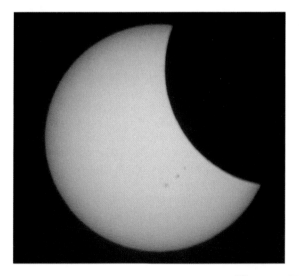

▶ This image of the partial phase shows quite a substantial chunk of Sun being covered by the lunar disk.

◄ When photospheric
light shines through the
last, deepest lunar valley,
a beautiful effect called
the diamond ring is
briefly visible.

the time of the eclipse will tell you which part of the Moon will be silhouetted at its limb against the Sun, and if you have a good Moon
map you can check beforehand which mountains and valleys will be
visible. This is useful for predicting effects like Baily's beads and the
diamond ring.

About 40 to 20 seconds before totality, as second contact
approaches, the peaks of lunar mountains will reach the edge of the
solar disk, and sunlight will shine through the lunar valleys. Bright
beads of light can sometimes be seen along the edge of the Moon in
the final seconds before the entire photosphere is covered and before
the ghostly light of the Sun's outer atmosphere, the corona, snaps
into view.

The beads of photospheric light are called Baily's beads after the
English astronomer Francis Baily, who first described them as resem-
bling a string of beads at the eclipse on May 15, 1836. Quite a lot of
information about the surface of the Moon and the Sun–Earth–Moon
orbital geometry can be gleaned from observing Baily's beads: the
precise part of the Moon silhouetted against the Sun's limb can be
determined, and the depths and heights of the valleys and mountains
can be calculated.

Sometimes the beads will wink out one by one, leaving just one
shaft of light shining through the last, deepest valley. This is the diamond ring, and it is a wonderful though fleeting sight, seen just before

second contact as the photosphere is covered and just after third contact as the photosphere starts to reappear.

At an eclipse, the sight of a diamond ring at the start of second contact is usually greeted with a cheer, as the long wait through the partial stage has finished and totality is just about to start. In contrast, when it appears at the beginning of third contact, it is greeted with sighs, moans and groans because it indicates that totality is over.

Second contact

If the eclipse is going to be total, and an observer is located at a point on the Earth's surface over which the umbral shadow will pass, the observer will experience second contact, which is the point when the last of the photospheric light blinks out. As soon as this happens, the tenuous, ephemeral outer atmosphere of the Sun, the corona, springs into view. This is amazing to watch – even understanding what is happening will not prepare you for the slight feeling of anxiety that greets you the first time you see this happening. It is one time when you realize how powerful are the forces of nature.

As soon as the last of the photospheric light disappears, it is safe to look directly at the Sun. Despite being much hotter than the photosphere, the corona is very tenuous and does not have the power to destroy your eyesight.

Around the rim of the Moon you may see glimpses of red. This is the Sun's chromosphere, the region between the photosphere and the corona. How much of the chromosphere is visible depends on the stage

▶ At second contact, the lunar eastern limb reaches the Sun's eastern limb and Baily's beads and the diamond ring disappear. The chromosphere and any prominences will become visible.

◄ An eclipse at solar minimum. The corona shines out all around the black disk of the Moon.

◄ An eclipse at solar maximum. The corona is restricted to the equatorial regions by the strong magnetic field.

of the solar cycle (see Chapter 1). At times of solar maximum, there is more activity and hence there should be more chromospheric activity visible. Sometimes it is possible to see the reddish glow of prominences erupting from the Sun's limb.

The ghostly corona

It is the corona that dominates during totality. The corona stretches its silvery white tendrils into space around the black disk of the Moon. If the Sun is near solar maximum, the corona will be contained by the increased magnetic field to zones along the Sun's equator. At solar minimum, the corona will be more spread out all around the Moon's disk.

If you have any form of magnification, even if only binoculars or a camera lens, use it to take a good look at the corona. It is perfectly safe and no filters are needed – just be careful to look away immediately totality is over and the photosphere reappears. The corona is composed of individual streamers which can extend out to ten solar diameters from the Sun. Some appear twisted. Shorter streamers, known as plumes, are emitted from the solar polar regions.

During totality the brighter stars and planets can become visible. It was during a solar eclipse that one of the predictions of Einstein's general theory of relativity was successfully verified (see the feature on pages 102–103). Occasionally a comet passing near the Sun can flicker into view during totality. This has happened four times during

▶ If the Moon is not quite at the correct distance for its disk to fully cover the Sun, a ring of photosphere known as the "ring of fire" can be seen and the corona will not become visible. An annular eclipse results.

modern solar eclipses: Comet Tewfik appeared during the May 17, 1882, eclipse; Comet Rondanina–Bester was visible at the May 20, 1947, eclipse; Comet C/1948 V1 was seen during the eclipse of November 1, 1948; and Comet Hale–Bopp could be observed at the eclipse of March 9, 1997.

The darkness during totality is not like ordinary night because of the ghostly light from the corona. Although the level of light from the corona is about the same as from a full moon, it is a different kind of light: eerie but beautiful. The darkness is also different because it falls so suddenly; ordinary night falls slowly, giving us time to adjust to the falling light levels.

For annular eclipses, second contact is defined as the moment when the ring phenomenon appears, with the Moon's western limb, or right-hand edge as it appears in the sky, just touching the inside of the Sun's western or right-hand edge.

Third contact

No matter how long totality lasts, it still seems to go in a flash. All too soon, the photosphere starts to peep through the lunar valleys at the western limb. As soon as the merest hint of photospheric light reappears, the shimmering coronal light vanishes. Totality is over, and observers should not look directly at the Sun again except through proper filters. If you can remember, just after totality look toward the east and you may see the Moon's shadow receding rapidly away from you along the ground.

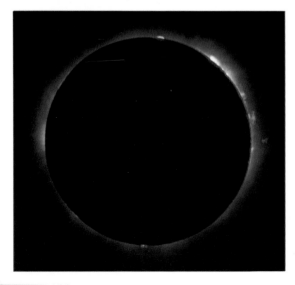

◀ Third contact occurs when the Moon's western limb starts to move away from the Sun's western limb and the photosphere reappears.

▶ As the final part of the Moon leaves the solar disk, fourth contact occurs and the solar eclipse is over.

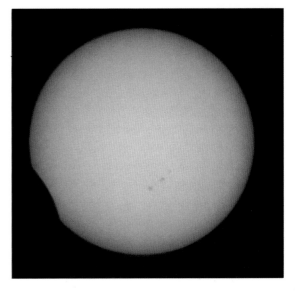

It does not take long for normality to return. The light brightens back to daylight levels, the temperature rises, and birds and animals return to their daytime pursuits.

For total eclipses, third contact occurs when the photosphere reappears. For annular eclipses, third contact is defined as the moment when the ring stops being complete, when the Moon's eastern limb, or left-hand edge as it appears in the sky, just touches the inside of the Sun's eastern or left-hand edge.

Fourth contact

Slowly, the Moon will leave the disk of the Sun, and the eclipse enters a reverse partial phase as the penumbral shadow moves away from the observer. For amateur observers, this final phase is somehow not so interesting as the partial phase before totality, when there was the rising anticipation of the approach of totality. Many do not bother to observe, packing away their equipment and discussing the spectacle that has just ended. Fourth contact occurs when the last of the Moon leaves the disk of the Sun.

Experiencing eclipses

As a partial eclipse approaches the 80% mark, it will become obvious that there is something happening. It is noticeable that there is a drop in sunlight, similar to a cloud crossing the Sun, though there is no cloud in sight. With the partial phase approaching 90%, there starts to be a fall in temperature. If you are observing from a hot climate, the contrast in temperature is more noticeable. It is an unreal experience to feel this drop in temperature with no clouds around to cause it in the normal way.

Depending on your place and country of observation, winds can start to pick up as the temperature drops. Very frustratingly, clouds can even form out of an apparently clear sky just as totality begins. Events occur just as they do when the Sun sets, but much more quickly – which is the strangest feeling, especially if the eclipse is early in the morning.

Verifying Einstein's theory of general relativity

Albert Einstein published his theory of general relativity in 1915. He effectively removed the need for the force of gravity, explaining gravity as the distortion of space-time. We perceive the Universe as three-dimensional (up, along and sideways), but Einstein proposed that it is four-dimensional, with time as the fourth dimension. If something very heavy sits in space-time, the space-time is distorted, a bit like a heavy ball being placed on a taut rubber sheet, causing the rubber sheet to dip slightly where the ball is sitting.

Einstein argued that light (or any other type of radiation) would be bent by a massive object because of the distortion of space-time. Going back to the ball on the rubber sheet, if you roll a marble close to the ball on the sheet, it will not roll in a straight line, but will move in and out of the dip in the sheet caused by the weight of the heavy ball, and its path will be deflected as a result. A beam of light passing near a massive object such as the Sun will move in and out of the warped space-time in a similar

way: the path of the light passing close to the Sun is bent.

Thus the light from a star, lying a long way from the Sun but close to the Sun in the sky, is bent as it passes close to the Sun. This causes the star to be seen at a slightly different place from normal. A star observed near the Sun's limb would be displaced by about 1.75".

In the days before coronagraphs (see Chapter 2), the only time it was possible to view stars close to the solar limb was during a total solar eclipse.

In August 1914, German scientists traveled to view the eclipse of August 21, but were arrested on Russian territory. They were held captive for about a month, and did not succeed in photographing the eclipse.

A more successful expedition occurred in 1919, when the English astronomer Arthur Eddington arranged for the eclipse of May 29 to be observed from two sites, one in Brazil and one from the island of Principe off the coast of West Africa. The star cluster, the Hyades, would

Seeing the totality of a solar eclipse is an unforgettable experience. No words or photographs can convey the eeriness of the coronal light or the feeling of the power of nature as the Moon marches inexorably across the Sun's disk. It is very emotional, and some of the sternest astronomers have been reduced to tears by the beauty of totality.

The Moon's shadow

The Moon's shadow races across the Earth's surface at speeds between 1800 and 7200 km/h. If you are at a suitable observing site, look toward the west just as the last of the photosphere is gobbled up by the Moon. Seeing the blackness race toward you is awe-inspiring: the shadow seems to be a tangible object, and there is nothing you can do to stop its progress toward you. The only place the shadow does not move west to east across the Earth's surface is at the poles.

Just before the end of totality, look down toward the horizon. If you have a good horizon, you will see a bright rim like a sunrise expanding rapidly as the end of the shadow rushes toward you. If you have time to remember, look round the horizon during mid-eclipse. In theory you should be able to see a 360° sunrise due to the light from the chromosphere and photosphere shining around the Moon's shadow which covers you.

A minute or so before totality begins and ends, remember to glance down at your feet, particularly if you have brought a white sheet or card to spread on the ground. You may be lucky enough to see shadow bands. Shadow bands are mottled, wavering patterns of dark and light which briefly cross the ground at this time; they are thought to be

be lying close to the limb of the Sun. The results verified Einstein's theory: the photographs taken showed the Hyades displaced by between 1.65″ and 1.98″.

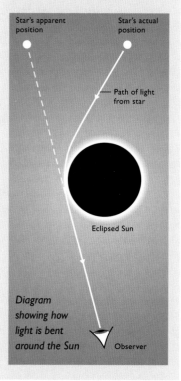

Star's apparent position

Star's actual position

Path of light from star

Eclipsed Sun

Observer

Diagram showing how light is bent around the Sun

caused by irregular refraction of the thin crescent Sun's light through the Earth's atmosphere. Shadow bands look something like the ripples seen at the bottom of swimming pool.

Myths and legends

If you did not know in advance that a partial solar eclipse was due, it could occur without you being aware of it. A reduction in the Sun's light only starts to become noticeable as the Moon covers more than 50% of the Sun's photosphere. With today's world of indoor lights and tall buildings, it would be an interesting experiment to see if anyone going about their everyday business noticed anything unusual, even with half the Sun's disk covered.

For people much more attuned to nature and much less cushioned against the elements, a solar eclipse was something quite frightening. Many myths and legends grew up in ancient times to account for a solar eclipse.

The first recorded account of a solar eclipse comes from India, where it is mentioned in a chronicle. The date was October 21, 3784 BC. Eclipses in ancient India were said to be caused by two dragons, one at each node, trying to eat the Sun.

Another old record of an eclipse comes from ancient China, in 2136 BC, where they also believed that the event was caused by a large dragon swallowing the Sun. Court astrologers were assigned the task of predicting further such terrifying events, but with the astronomical knowledge of the time it was a difficult task. Some sources state that the court astrologers Hsi and Ho were executed for failing to predict another eclipse because "they were so sunk in wine and excess that they neglected the ordering of the seasons and allowed the days to get into confusion." The ancient Chinese would bang drums and send up fireworks in order to scare the dragon away.

Eating the Sun

Other early cultures also believed that eclipses were caused by various animals eating or swallowing the Sun. The Mandarin word for eclipse is "shi," which actually means "eat." In Indonesia, they too thought a dragon was trying to devour the Sun. In Siberia it was a vampire; in Bolivia, a huge dog.

The early Vietnamese believed that a frog was trying to swallow the Sun. At other times the frog would be chained and kept in a pool by the lord Hanh. When it escaped it would chase the Sun, but the ladies of the Moon would waken Hanh, who was the only person able to make the frog disgorge the Sun. Today in Vietnam, people still beat their rice bowls during an eclipse to help the ladies rouse Hanh from his slumber.

Later Chinese cultures explained solar eclipses as being caused by the Sun and Moon having different Yin and Yang rhythms of brightness and darkness, Yin and Yang being the Earth's two fundamental forces.

A clay tablet discovered during excavations of the ancient Mesopotamian town of Ugarit describes the eclipse of May 3, 1375 BC. The account reads: "On a day of the new Moon, in the month of Hiyar, the Sun was ashamed, and hid itself in the day-time, with Mars as witness."

War eclipses

On at least two occasions, wars have been affected by the occurrence of a solar eclipse: on one occasion ending a conflict, on another, being the cause of conflict.

On May 28, 585 BC, the six-year battle between King Alyattes of the Lydians and King Cyaxares of Medes ended when the eclipse occurred, frightening both sides so much that they agreed a peace in order not to further offend the gods.

In AD 840, Louis Le Debonnaire, who was son and heir to Charlemagne's huge empire, is reputed to have died of fright during a solar eclipse. His three sons subsequently waged war over the kingdom, which was eventually divided into France, Germany and Italy.

Science from eclipses

Before the invention of the coronagraph, and the development of space technology allowing observations to be carried out above the Earth's atmosphere, the only time the corona could be studied was during solar eclipses. It was not until the late nineteenth century that a connection was firmly established between the Sun and the halo of the corona.

The corona was originally thought to be a halo caused by earthly fumes. In 1567, Johannes Kepler attributed it to the Moon, thinking it was a lunar atmosphere refracting rays from the Sun and hence shining by this light. Solar prominences were thought to be clouds in the lunar atmosphere.

The first successful solar eclipse photograph was taken in 1851, showing the inner corona and prominences. At an eclipse visible from Spain in 1860, two astronomers, Warren De La Rue and Angelo Secchi, photographed prominences from sites that were 500 km apart. The resultant photographs showed no shift in the positions of the prominences. If they had been only at the distance of the Moon, then a change in position was expected. It was thus deduced that the prominences were associated with the Sun. Subsequent eclipses were viewed spectroscopically, and the nature of the spectra obtained from prominences and the corona showed that they were indisputably solar in origin.

New elements

One of the leaders of solar eclipse spectroscopy was Jules Janssen. During the eclipse of August 18, 1868, which was visible from India, Janssen noticed a bright yellow emission line near the sodium D lines at wavelengths of 589.0 and 589.6 nm in the spectra of prominences (see Chapter 1). Also noticed later by the Englishman Norman Lockyer, it was thought to be a line indicating the presence of a new element, as no known terrestrial element produced an emission line at that wavelength. It was named helium (after the Greek Sun god). Thirty years later, helium was discovered on Earth by William Ramsay, in gas released from a certain type of rock.

Janssen was an astronomer who caught the eclipse bug badly. In addition to it being the only time he could observe the corona and solar prominences, he also enjoyed the spectacle. For the December 11, 1870, eclipse visible from the Mediterranean, Janssen was trapped in Paris, then surrounded by Bismarck's Prussian troops. Determined to see the eclipse, Janssen loaded his equipment into a balloon and flew over the besieging troops. Although he managed to reach northern Algeria, which was under the path of totality, it was clouded out – a fate not unknown to present-day eclipse observers.

Another eclipse addict, Dmitri Mendeleev (who devised the periodic table), was determined to get above any clouds during the August 19, 1887, eclipse visible from Russia. He also went up in a balloon and ascended to 3.5 km.

During the eclipse of August 7, 1869, Charles Young and William Harkness discovered an emission line in the inner corona. Again it lay at a wavelength where no known terrestrial element produced emission lines: this time in the green, at a wavelength of 530.3 nm. It was named coronium, and several more emission lines with wavelengths not known on Earth were subsequently discovered. There was a problem, however, with fitting the new elements into the periodic table. Eventually, in 1939, it was realized that the coronium line and the other emission lines in the corona's spectrum were from elements that were very highly ionized, implying that the corona was at a temperature of some 500,000 K. The line originally attributed to coronium is actually from iron ionized 13 times.

Later eclipses

Early in the twentieth century, an attempt was made during an eclipse to verify Einstein's general theory of relativity (see pages 102–103). As the twentieth century progressed, astronomers built on the observations of the early eclipse observers, determining the nature of solar prominences, the chromosphere and the corona. Photography became more sophisticated, and different exposure times were used to record different phenomena within the corona. The nature of the corona was widely debated.

Spectroscopy and the method of recording spectra for further analysis was perfected, giving insight into prominences and the chromosphere. The Frenchman Bernard Lyot, after whom the Lyot filter is named, took a leading role in developing coronagraphs, filters and polarimeters. Observing eclipses in the 1940s, he developed instruments to record coronal lines across the whole of the visible spectrum. He was successful during the February 25, 1952, eclipse, which he observed from Khartoum, Sudan. This was despite all his instruments suffering during a sandstorm three days earlier, after which he had to dismantle and clean every one.

Despite being successful, the Egyptian authorities would not allow Lyot to keep his results. They hoped he had made a major discovery, and wanted to share in any recognition from the international scientific community. Sadly Lyot died of a heart attack on April 2, 1952, and it took his widow over two weeks to repatriate his body.

Eclipse photography continued to be high on the agenda until the 1980s. Only with the advent of solar spacecraft such as SOHO (see Chapter 9) has the scientific significance of observing eclipses diminished, but it is still a time when aspects of both the Sun and Moon can be seen from the ground that cannot be seen at any other time.

7 · SNAPPING THE SUN

Photographing the Sun is a challenge because of its dangerous brightness. Apart from sunset and sunrise, when most of the Sun's light is below the horizon, it is not possible to just point an ordinary camera at the Sun and take a snapshot. Indeed, even at sunrise and sunset the greatest care should be taken if photographing the Sun directly: just a glimpse of the photosphere through the camera lens or viewfinder is enough to damage the eyes. It is advisable never to point a camera directly at the Sun, even if it is low in the sky.

Solar photography can be done either by photographing the projected image (see Chapter 3 on how to project the Sun) or by photographing directly through filters (see Chapter 2 for information on filters).

Choice of camera

Most people today own a camera, and you can probably use the camera you have to take pictures of the projected Sun's image, although manually operated single-lens reflex (SLR) cameras are the best type to use. To take direct pictures of the Sun you will need filters, and generally a 35 mm SLR camera is best for this. It is possible to hold a filter up to an ordinary "snapshot" camera and take a picture of the Sun, but this can be dangerous and the result is unlikely to be worth the risk. I strongly recommend that if you want to experiment with photographing the Sun with filters, you use a camera to which the filter can be properly attached.

Some SLR cameras are very sophisticated, with features such as built-in zoom lenses, automatic focus and flash, and are good for everyday pictures, but the automatic functions become a liability when doing solar photography. For photographing the Sun (indeed, for any astronomical photography), the simplest SLR camera is the best. Many older second-hand cameras are ideal for astronomical photography. The main requirement is that the camera allows manual selection of aperture and exposure time. Also ensure that it has a standard lens fitting, so that you can attach cheap, readily available accessories.

Since the Sun is very bright, choose a camera on which you can select suitably short exposure times. Times of 1/125 second down to 1/2000 second are ideally needed, and the minimum requirement is at least down to 1/500 second.

The 35 mm format camera is the most widely used, though other formats do exist, but obtaining and developing film for cameras other than 35 mm can be difficult.

A steady base

Your camera needs to be stable when taking pictures: any movement as the shutter opens will blur the image. This becomes more of a problem when using larger lenses: not only is the camera heavier to hold, but also the increased magnification exaggerates any camera movement. A sturdy tripod is a good investment.

Check that the shutter does not cause the camera to vibrate when used. At a little expense, a camera release can be bought which will allow you to open and close the camera shutter without actually touching the camera, thereby minimizing camera shake.

Screens and adapters

A few SLR cameras have removable focusing screens, which are ideal for astronomical photography because different screens can be selected for different objects. For the Sun, split-field screens and microprisms are not easy to use because they break the image in the viewfinder into bright and dark spots, which are impossible to use for focusing. Clear screens are better, but ground glass screens with a clear central area have proved to be the best for both direct solar photography and for photographing a projected image.

Older cameras generally have an adapter that allows them to be fixed to a tripod, which is useful if the camera is to be used with its own lens to photograph the Sun. If you want to photograph through a telescope, then a suitable adapter is needed. For ordinary SLR cameras these adapters are not hard to obtain, but there are some more unusual second-hand camera bodies around with strange lens fittings. If your camera is not standard, it can be difficult and expensive to buy different lenses or to attach the camera to telescopes.

Choice of film

If you have a 35 mm camera, then any film bought in the high street can be used to take pictures of the Sun, particularly if you are photographing its projected image. The lower the film speed (the ISO number), the better, because it has a finer grain and so gives a sharper image. The specialist should look for a high-resolution film that is sensitive over a wide spectral range but has low sensitivity.

Photographing a projected image

Photographing a projected image is the easiest and safest method of photographing the Sun. It is also the cheapest because in theory you need no special equipment: you can use your normal camera. In effect all you are doing is photographing a sheet of paper with the Sun's image on it.

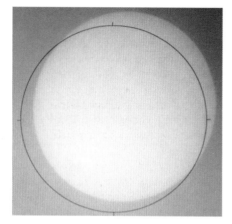

▲ It is possible to photograph a projected image of the Sun, though experimentation with getting the optimum exposure time and set-up of equipment will be needed.

▲ In this projected image of the Sun, the solar disk is not completely circular, showing that the image is projected at a slight angle to the screen.

In practice, though, there is more to it than just taking a few shots of the projected image. The image is usually very faint, so you need to use the shortest possible exposure time. To enhance the contrast of the sunspots against the white card or paper, you need to shield the image from the direct rays of the Sun – and that washes out the image. The darker the surroundings of the image, the more contrast you will see, but the trade-off is that you will need slightly longer exposure times.

If you manage to get the surroundings to the image very dark, you will increase the contrast, but some modern cameras have built-in flashes which are difficult, if not impossible, to switch off. In the dark surroundings, the flash will activate, completely washing out your solar image. One way round this is to tape over the flash on the camera with black tape.

Experiment with your camera and projection box as it is very possible to get good images of sunspots using just an ordinary non-SLR camera. I project the solar image through my telescope into a cardboard observing box (see Chapter 3), and for photography with my ordinary camera, I shield the camera and the majority of the path of the image rays with a cloth, allowing just a small amount of light in and so stopping the camera flash activating. This method takes time to perfect, but once the set-up for your equipment has been streamlined, it is a quick and easy way of obtaining quite pleasing photographs of the Sun's disk.

Direct photography with filters

Just as direct observation of the Sun must be done with extreme care, so must direct photography. If anything, the dangers are greater because you may be staring through your viewfinder for longer, concentrating on focusing the image. Safe, correctly installed filters will allow you to look directly at the Sun for the length of time needed to focus an image, and the danger is minimized. So, as with direct observing (see Chapter 2), always ensure that any filter you use is from a reputable dealer, fixes to the objective end of the telescope or camera lens, is correctly fitted and is in good condition.

Direct photography can be done either with just the camera lens or by using a telescope as the lens. It is also possible to attach your camera to specialized instruments such as spectroscopes or hydrogen-alpha filters and hence to photograph the Sun at wavelengths other than those of white light.

If you use just your camera with an ordinary lens (the focal length of a standard lens on an SLR camera is usually between 45 and 55 mm), the image of the Sun on your photograph will be small. Although the Sun is exceedingly bright, it is the same size as the Moon in the sky, 0.5° across. The image size will be the same as that of a 200 mm coin photographed from about 2 meters away. To get a larger image, you need a more powerful lens (see below).

Experimenting with exposures

The exposure time for direct viewing of the Sun will depend on many factors, for example lens, filter and film, and a complete discussion of this is outside the scope of this book. With patience, however, it is quite easy to work out the best exposure time for your equipment by bracketing your exposures (see below). At worst, this method will cost you a bit of time and some rolls of film, but after experimenting you should be able to judge the correct exposure time much more accurately, thereby minimizing subsequent waste of film and the disappointment of poor images.

Select the equipment you want to use, making careful note of everything so that you can reproduce the exact configuration later. Using the film of your choice, take a range of exposures, making careful note of the time for each exposure. Then, when you have the film developed, you can see which is the best exposure time.

Maximum exposure times

The exposure time will depend on the focal length of camera lens or telescope. The focal length of a lens is the distance from the lens to the point where the image is focused. For a refracting telescope, the focal

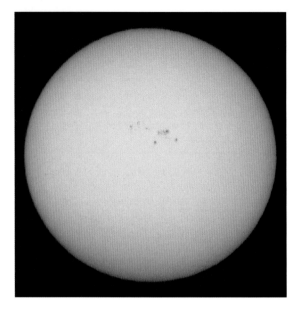

◀ Details of the Sun can be recorded by directly photographing the Sun through a filter. A Mylar filter will give a blue-tinged image.

length is the distance from the objective lens to the point of focus; for a reflecting telescope it is the distance from the primary mirror to the point of focus. As a guide, for a given focal length of lens or telescope, the maximum exposure time before the image of the Sun will start to trail across the photograph because of the Earth's rotation is

$$t = 500/f$$

where t is the exposure time in seconds and f is the focal length of lens or telescope in millimeters.

For example, with a 200 mm lens the maximum exposure time before the image trails is $500/200 = 2.5$ seconds. Given the brightness of the Sun, however, unless you have a very dark filter you will be using exposure times much shorter than this. Use the suggested exposure times for the partial phase of a solar eclipse as a guide (see the tables at the end of this chapter).

If you are using a specialized filter, such as a Lyot coronagraph or hydrogen-alpha filter, try bracketing exposures between 1/15 and 1/125 second.

Camera lenses

Standard lenses for SLR cameras (45 to 55 mm) produce a small image of the Sun with little detail. For more detail and larger images, you need a telephoto lens or a small telescope. Telephoto lenses come in a broad range of focal lengths, and can be quite expensive to buy. Many lenses are available second-hand, but unless you are absolutely sure that the lens you intend to buy is not damaged, I would recommend buying new. The lens is the most important part of the imaging system, and the part that is the most easily damaged. You cannot always tell whether a lens is damaged just by looking at it.

To see detail such as sunspots, lenses of focal length 400 mm are needed. When photographing at this magnification, a sturdy tripod and camera release are essential.

Keeping a record

Do not forget to make a record of the time and date of your photographs, and note the number of the frame on the film for each exposure. It is quite easy to forget this information, and when you get your developed film back it can be difficult to work out which image is which. The more thorough your note-taking, the easier it is to interpret your results. For more serious work, also note the weather conditions since a slight haze or a wind can have a large effect on the quality of the photograph.

It is also useful to record the precise equipment used, together with the exposure time. This information will help you to work out the best equipment configuration and the best exposure times for each configuration. When experimenting, bracket your exposures: that is, take pictures with a range of exposure times so that when you have them developed you can see which time was best for the equipment you were using and the prevailing conditions. As you become more experienced, your bracketing can be reduced, but it always a good idea to bracket at least one time interval each side of the exposure time you expect to be the best, as atmospheric conditions can change on a timescale of minutes.

Developing your own film is very satisfying, but it is a another specialized skill to learn, and the equipment can be expensive. If you take your film into a high-street store to be developed, always let them know that the film contains some astronomical images. This will ensure they do not neglect to develop them, believing them to be the result of errors in either film or camera. I always ask that they develop the entire film. Also, take a "normal" photograph before and after your solar photography so that the developer can ascertain the correct levels of color and contrast for development.

Photographing an eclipse

Just experiencing an eclipse is magical, but in today's world people want to have a photographic record of such experiences. I would strongly recommend that at your first total eclipse you spend most of your time simply watching events unfold. The problem with photography is that it can use up all those precious minutes of totality, so that when the event is over the observer has not observed at all.

As with photographing the everyday Sun, there are many ways in which to photograph an eclipse. If you are photographing an eclipse for the first time, I suggest that you follow a very modest photographic plan to allow yourself plenty of time just to look around and absorb the events and the atmosphere. At the end of this chapter is a simple plan for photographing the Sun directly. Remember that no special filters are needed during totality.

Photographic equipment

Just as with photographing the non-eclipsing Sun, your photographic program will depend on the equipment you have. If you decide to upgrade your equipment especially for the eclipse, familiarize yourself with it all beforehand. Excitement runs high during an eclipse, time is short during totality, and you have to deal very quickly with fluctuating light levels. Practice with your equipment until you have your procedures down to a fine art. This will minimize mistakes and allow you time to look around as well.

An ordinary snapshot camera will allow you to take pictures of your fellow eclipse-watchers, the partial phase in projection, and phenomena such as shadow bands, partial Sun-shadows, the onset of the Moon's shadow and possibly an atmospheric totality shot. Be warned, however, that you will be very unpopular if your automatic flash goes off during totality, spoiling everyone else's photographs. Ensure that you have disabled the flash unit or taped over it very securely.

If you have a 35 mm SLR camera with a standard lens, in addition to everything you can photograph with an ordinary snapshot camera you should be able to take successful images of totality because you will be able to vary the exposure time manually at this stage of the eclipse. The Sun's image will not appear very large on your film, but careful bracketing of exposure times could allow you to capture any bright stars, planets or even comets in the vicinity of the eclipsed Sun.

As with any type of photography, stabilizing your camera is very important, especially if you intend to take longer exposures at totality. For this, a good tripod is essential, and the use of a camera release is strongly advised in order to minimize camera shake.

More powerful lenses

Obviously, the more powerful the lens you have fixed to your camera, the more detail in the corona you will be able to photograph at totality, and if you have a filter to fit to the end of your camera lens you will be able to photograph the partial phase directly. Unless you have two cameras, one for the partial phase and one for totality, any filter must be easy and quick to remove and replace. The time between the onset of Baily's beads or the diamond ring and totality is a fraction of a second. In this time you need to remove the filter, and more importantly to replace the filter as the photosphere starts to reappear at the end of totality.

If you intend to use a camera fixed to a telescope, a smaller rather than a larger instrument is advised. One reason is that you are likely to have to travel to the eclipse site, so obviously the smaller and more compact your equipment, the better. Another reason is simply that you do not need a large telescope. In fact if it is too large, you may have trouble with the seeing. Since you are, by definition, observing during the day, the atmosphere will probably be quite turbulent. During an eclipse the temperature fluctuates quite dramatically, increasing atmospheric turbulence, which causes blurring in photographs. The larger the telescope or lens, the greater the effect of poor seeing. Not many eclipse photographers use telescopes more than 200 mm in aperture.

Capturing the surroundings

For most of the partial phase, photographing your surroundings will be the same as in ordinary daylight. Once you are all set up, the partial phase usually lasts long enough to give you time to capture what is going on around you. Some very atmospheric photographs can be obtained of your surroundings and fellow eclipse-watchers, particularly if you have traveled to an exotic location to watch the eclipse. I have watched eclipses from an Indian graveyard, a Caribbean beach and a caravan park in France!

When you photograph totality, try to have an interesting foreground, particularly if you are using small lenses. The eclipsed Sun might appear small, but it can make an interesting and atmospheric shot. Such shots are enhanced by using wide-angle and fish-eye lenses. With wide-angle lens (20–40 mm) you can frame the Sun with some landscape feature – either natural, like mountains or trees, or artificial, such as buildings or statues. A fish-eye lens (less than 20 mm) allows you to bring in more of your surroundings. The Sun will appear small on your image, but you may be able to image the false sunrise effect around the horizon.

Multiple exposures

If you have a camera that allows manual film advance, then you can take a sequence of shots without moving the film on, and can capture all the phases of the eclipse on one photograph. The first thing you need to do is to work out where to point the camera so that it is in the right direction for all the shots. Decide whether you want the totally

◀ *The annular eclipse seen from California on January 4, 1992, has been captured accurately, but the atmosphere of the event has also been recorded, with excellent use of the surroundings and the hot air balloon.*

eclipsed Sun in the middle of your frame or at one edge, then work out this position and hence the angle of your camera. It may help to go to your eclipse site the day before so you can actually see where the Sun is going to be in the sky at the time of the eclipse.

Plan well beforehand – make a list of exposure times for each phase, and take a picture at regular intervals (between 2 and 10 minutes apart). Many people make good use of a small tape recorder for doing multiple exposure photographs. Plan what you are going to do, then record, in real time, commands for yourself. For example, if you intend to take shots 5 minutes apart, record messages to yourself 5 minutes apart on the tape to warn yourself that the time is approaching for the next shot and remind yourself of exposure times. These multiple exposures require a lot of planning, but are very effective and satisfying to do.

A simple photographic plan

If this is your first eclipse, do not plan to spend a great deal of time behind the camera lens at totality, just enjoy the experience. There will still be plenty of time during the partial phase to take some interesting shots. Plan ahead: perhaps take a photograph every 5 minutes during the partial phase, or an image of the Sun every 10 minutes interspersed with scenic shots and images of events such as the falling light levels or the partial Sun-shadows.

For totality, the simpler the photographic plan, the less there is to go wrong and the more time you can spend just gazing at the spectacle. Make sure that your filter is easily removable (but not so easy that it falls off during the partial phase!). Plan where you will keep the filter so that you do not lose it or step on it by accident, and it is easily accessible for when totality is over. Alternatively, if you are lucky enough to have two cameras then use one for the partial phase and one for totality.

For a first eclipse I suggest trying three exposures for Baily's beads or the diamond ring, and perhaps five or six different exposures during totality. It is very easy to get hooked on trying to photograph the

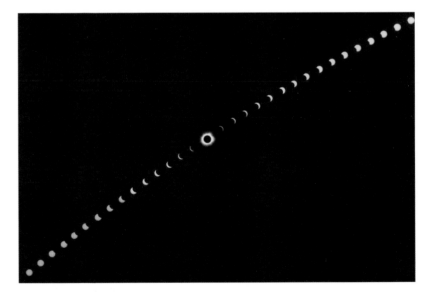

▲ A time lapse image of a solar eclipse will give an idea of every stage throughout the eclipse. In this image both partial phases have been photographed. Sometimes people just capture one of the partial phases, with the totally eclipsed Sun at the side of the image.

eclipse, but please do not forget to look directly at the eclipsed Sun, not through the lens, and to look around to get a feel of the occasion. After viewing your first total eclipse, you will understand why people travel halfway round the world to inhospitable places, just to experience a few more minutes of totality.

The tables that follow give a guide to exposure times at various points during the eclipse. They are for direct photography, so the times will vary slightly according to the type of filter you are using. Experiment beforehand with your equipment to work out the best exposure times for your filter and lens. Also practice with your equipment until everything is second nature. Eclipses happen quickly: there are rapidly varying light levels, and this, coupled with a high level of excitement, makes it is easy to make mistakes, to forget something or just to get so wrapped up in taking photographs or struggling with equipment that the eclipse happens without you experiencing it.

Make sure you have plenty of film and that you will not need to change films just as totality starts. Any batteries you need for any equipment should be fully charged, spare ones should be to hand. Do not forget to remove your filter during totality!

The following tables can be used as a guide for photographing solar eclipses with a 35 mm manual SLR camera. These times are only guides, because the equipment, location, atmospheric conditions, filter used for the partial phase, and so on, will all help to determine the best exposure times. The percentage (magnitude) of the partial phase will affect the exposure time too: the greater the phase, the longer the exposure required. Whatever stage of the eclipse you are photographing, always bracket across the suggested exposure times.

In the following tables the focal ratio is f/D, where f is the focal length and D is the aperture. All exposure times are in seconds. Each table is for a different film speed (ISO number).

FOR ISO 100 FILM:						
Focal ratio	Partial phase	Baily's beads	Chromosphere	Inner corona	Outer corona	Surroundings
2.8	—	—	—	1/60	1/30	1/2
4	—	—	1/1000	1/30	1/15	1
5.6	—	—	1/500	1/15	1/8	2
8	—	1/2000	1/250	1/8	1/4	4
11	1/2000	1/1000	1/125	1/4	1/2	8
16	1/1000	1/500	1/60	1/2	1	15
22	1/500	1/250	1/30	1	2	30
32	1/250	1/125	1/15	2	5	—
44	1/125	1/60	1/8	5	10	—

FOR ISO 200 FILM:						
Focal ratio	Partial phase	Baily's beads	Chromosphere	Inner corona	Outer corona	Surroundings
4	—	—	—	1/60	1/30	1/2
5.6	—	1/2000	1/1000	1/30	1/15	1
8	—	1/1000	1/500	1/15	1/8	2
11	1/2000	1/500	1/250	1/8	1/4	4
16	1/1000	1/250	1/125	1/4	1/2	8
22	1/500	1/125	1/60	1/2	1	15
32	1/250	1/60	1/30	1	2	30
44	1/125	1/30	1/15	2	5	—
64	1/60	1/15	1/8	5	10	—

FOR ISO 400 FILM:						
Focal ratio	Partial phase	Baily's beads	Chromosphere	Inner corona	Outer corona	Surroundings
5.6	—	—	—	1/60	1/30	1/2
8	—	1/2000	1/1000	1/30	1/15	1
11	—	1/1000	1/500	1/15	1/8	2
16	1/2000	1/500	1/250	1/8	1/4	4
22	1/1000	1/250	1/125	1/4	1/2	8
32	1/500	1/125	1/60	1/2	1	15
44	1/250	1/60	1/30	1	2	30
64	1/125	1/30	1/15	2	5	—
88	1/60	1/15	1/8	5	10	—

— 8 · THE SUN–EARTH CONNECTION —

The Sun has a profound effect on the Earth, and on all other objects in the Solar System. Some of the effects are dramatic and obvious, some much more subtle, but all are important to life on Earth. Today, spacecraft and ground-based observatories monitor the Sun constantly; the more astronomers learn about the Sun, the more it is realized how important the solar–terrestrial connection is, and in some cases, how human behavior can upset the delicate balances.

Stating the obvious

The most obvious effect is night and day. The Earth rotates on its axis in a period of 23 hours 56.1 minutes, and it is this rotation that causes an observer to sometimes be on the side of the Earth pointing toward the Sun, experiencing day, and sometimes on the side pointing away and experiencing night. We use this rotation to define our "day." Our day is important to us as our natural living rhythms are dictated by the cycle of day and night. Modern living with its artificial lights may have divorced us a little from the celestial clocks, but few of us can go without some sleep during a 24-hour period.

For ease of timekeeping we define our day as 24 hours. The difference of 4 minutes between this and the actual rotation period may seem small, but over time it mounts up fairly quickly. This is the reason for adding a day to the year every 4 years. In addition, because the Earth's orbit is an ellipse (see below), its orbital rate varies a little over the course of its "year," and this has a small effect on the length of the day. Occasionally a leap second or two is added to keep our human clocks in time with the celestial mechanics.

The length of a terrestrial day is also not constant over time. It is slowly getting longer, because of the gravitational influence of the Moon. Over a century, the Earth's day gets longer by about 0.0014 seconds. During Palaeozoic times, the length of the day is thought to have been only 18 hours. In the future, the day could grow to be as long as 1320 hours (55 days of current length).

The Sun's gravitational tug

The Sun contains 99.87% of the mass of the Solar System, which is why all the bodies in the Solar System orbit it. The gravitational tug of the Sun causes the Earth to revolve around it in 365.26 days, giving us our year. The Earth's path is not a perfect circle with the Sun at the center; instead, the Earth's orbit is an ellipse with the Sun at one focus of the ellipse, so the distance of the Earth from the Sun varies through the year. At aphelion (the farthest point) the distance is about 152 million km, while at perihelion (the nearest point) they lie about 147 million km apart. Surprisingly, this difference in distance is not what causes the seasons.

The Sun's gravitational tug also has a small influence on the tides. The tides are caused mainly by the motion of the Moon around the Earth: since it is nearer than the Sun it has a greater effect despite being much less massive, but the Sun's influence still plays a part. When the Sun, Earth and Moon are aligned (at new and full Moon) the Sun's gravitational pull works with the Moon's and the rise in the oceans is greatest. These are the spring tides, and when the Sun lies over the equator, at the equinoxes, the highest spring tides occur. When the Sun, Earth and Moon lie at a right angle (when the Moon is at first or last quarter), the Sun's gravity works against the Moon's to produce the lower neap tide.

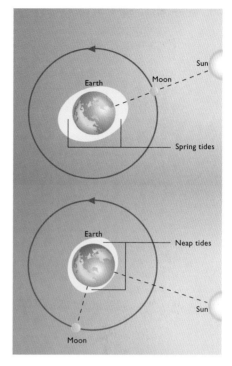

▲ Tides are caused by the gravitational tug of the Sun and the Moon. When they are aligned, their gravitational pull works together to give the highest tides, called the spring tides. When the Sun and Moon are at right angles, the lower neap tides occur.

A varying constant

The Sun is a dynamic object and is constantly changing: for example, its diameter fluctuates and the number of sunspots varies. Another changing factor is the total amount of energy it radiates. The Sun emits energy across the whole of the electromagnetic spectrum (see Chapter 1), and the amount of energy received at the Earth is known as the solar constant. However, this "constant" is not constant at all.

The exact definition of the solar constant is the amount of solar energy across all wavelengths received at the Earth above its absorbing atmosphere when the Earth is at its mean distance from the Sun (one astronomical unit) per unit area per unit time. This value varies, but it is around 1400 watts per square meter. It is sometimes known as solar irradiance.

The Sun and Earth's climate

The Sun's energy output has a great effect on temperatures across the Earth's surface and thus on the overall climate. Scientists have studied Earth's climate intensively as it is so important to the existence of life. A change in temperature of just a few degrees will bring about a marked change in climate and could have implications for the future of human life. Monitoring the value of the solar constant is therefore very important, but this is only one of many factors that affect the Earth's climate.

The Sun clearly has effects on the Earth's climate: at the very least, the changing angle of incidence of the Sun's light and heat falling on the Earth's surface over the course of the year causes the seasons (see the feature opposite). The Sun's heat also drives the wind flows in the upper atmosphere, producing vertical convection currents at low latitudes, which are known as Hadley cells. These cells are formed by hot air rising at the equator and then sinking, as it cools, at latitudes of about 30°N and 30°S. At higher latitudes, atmospheric circulation is mainly horizontal, with traveling anticyclones (areas of high pressure) and cyclones (areas of low pressure).

The Sun also drives the ocean currents, and these have a large influence on climate. In addition the Sun's energy interacts with gases in the Earth's upper atmosphere, which has an effect on the

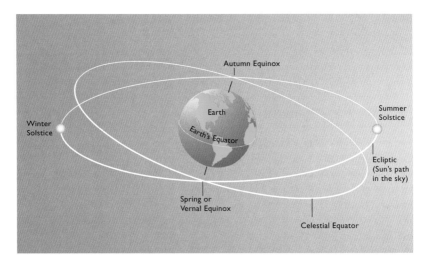

▲ The points where the Sun crosses the celestial equator are known as the equinoxes. At these times, the Earth experiences equal hours of day and night. At the summer and winter solstices, the Sun reaches the highest and lowest points, respectively, in the sky as viewed from Earth.

The seasons

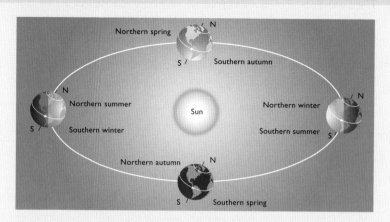

Many regions of the Earth experience seasons. Several factors dictate how much the temperature and weather vary during the year for each part of the globe. In general, countries in each hemisphere are warmer during their summer, and cooler in winter. This is not a result of the Earth's varying distance from the Sun – in fact, the northern hemisphere has its summer when the Earth is farthest from the Sun. The principle cause of the seasons is the inclination of the Earth's axis of rotation.

The Earth's axis is tilted at 23° 26′ to the ecliptic plane. The ecliptic plane is the plane in which the Earth orbits the Sun, so this tilt can be thought of as a tilt toward or away from the Sun. When the Earth is tilted toward the Sun, observers in the northern hemisphere will see the Sun higher in the sky. The Sun reaches the highest point in the sky over the northern hemisphere at the summer solstice.

The higher the Sun is in the sky, the less atmosphere its heat and light has to penetrate to reach the ground, hence the ground receives more energy and it heats up more. The Sun is also above the horizon for longer, which again means that more energy reaches the ground. These two factors have more effect on how much light and energy are received in a particular region on the Earth's surface than does the Earth's distance from the Sun.

For lucky observers in the southern hemisphere, summer falls when the Earth's south pole is tilted toward the Sun, which happens when the Earth is also nearest the Sun. In addition, the southern hemisphere has more of its surface area covered by ocean, and water has a greater capacity for storing solar energy than land. Southern-hemisphere summers are thus slightly warmer than those of the northern hemisphere, but their winters are cooler.

temperature at the surface of the Earth. There is no simple, overall description of how the Sun influences the Earth's climate, and a great deal of research is devoted to unraveling the many threads of a complex pattern.

A protective blanket

The Earth's atmosphere blankets the Earth, stretching up to heights of 1000 km, although at these altitudes it is extremely tenuous. Humans can survive without extra oxygen only in a very thin layer close to the Earth's surface. Even at the tops of the highest mountains, at heights of a mere 8 to 9 km, humans find it difficult to breathe. This blanket of gas – mainly nitrogen (78%) and oxygen (21%), with traces of other gases such as water vapor (up to 4%), argon (0.9%) and carbon dioxide (0.03%) – is finely balanced to give us a liveable environment. Small fluctuations in the solar output and in some human activities can change the composition of the atmosphere, triggering changes in the Earth's climate.

Without our atmosphere, life would not be possible on Earth, not only for the obvious reason that we would have nothing to breathe, but also because the atmosphere absorbs harmful solar radiation. We have to send instruments above the atmosphere to observe in the higher-energy part of the electromagnetic spectrum because X-rays and gamma rays do not reach the ground.

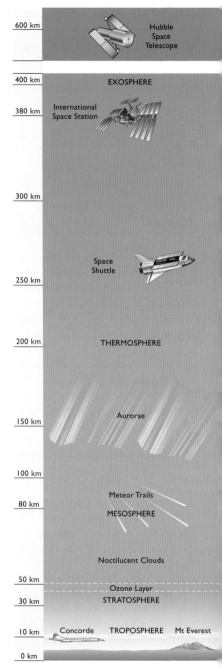

600 km — Hubble Space Telescope

400 km — EXOSPHERE

380 km — International Space Station

300 km

250 km — Space Shuttle

200 km — THERMOSPHERE

150 km — Aurorae

100 km

80 km — Meteor Trails

MESOSPHERE

Noctilucent Clouds

50 km — Ozone Layer

30 km — STRATOSPHERE

10 km — Concorde   TROPOSPHERE   Mt Everest

0 km

◀ *The Earth's atmosphere consists of the troposphere, extending from ground level to a height of between 8 and 17 km; the stratosphere, which extents up to about 50 km; the mesosphere, between 50 and about 100 km; the thermosphere, from about 100 up to 200 km; and beyond this the exosphere.*

The atmosphere also keeps the ground warm, like a true blanket. Without the atmosphere, the Earth's surface temperature would be about minus 45°C (228 K) – too cold to support higher life-forms. It is the greenhouse gases such as water vapor and carbon dioxide that help keep the Earth's surface at a temperature warm enough to support life.

Climate fluctuations

The Sun's energy output varies not only over the 11-year sunspot cycle but also over longer timescales. One of these longer-period

The greenhouse effect

The Earth's atmosphere acts like the glass in a greenhouse, letting the Sun's energy through but stopping some of the heat radiating back out again. The Sun's visible sunlight can easily pass through the atmosphere to the ground, where it is absorbed and re-radiated at slightly longer wavelengths. These longer wavelengths (in the infrared part of the electromagnetic spectrum) are then trapped by some of the constituents of the atmosphere, stopping the heat from radiating back out to space and so heating up the Earth's surface. The main constituents that stop the infrared rays from escaping are water vapor and carbon dioxide. Without this greenhouse effect, the Earth's surface would be about 30 K cooler.

Since the onset of the industrial revolution, the mean temperature over the surface of the Earth has risen by between 0.6 and 0.8 K per century. This temperature rise is connected with the increased amount of greenhouse gases emitted by human activity. In addition, the destruction of vast areas of vegetation has an affect on the amount of carbon dioxide in the atmosphere because plants and trees absorb carbon dioxide and emit oxygen during photosynthesis.

Unless these harmful emissions of greenhouse gases are halted, the global climate a hundred years from now is likely to be on average about 2 K warmer, and the effects on local weather systems could be major. There are already signs that storms are becoming more powerful because of the extra heat energy they contain. Another consequence of global warming is that the sea-level will rise (by perhaps half a meter in the coming century), not just because of ice melting in Antarctica and Greenland but also because of the thermal expansion of seawater as it warms up.

fluctuations may be the cause of the periodic ice ages suffered by the Earth through its history.

When the solar cycle was first discovered, astronomers looked at all manner of data for a connection with the Earth's climate. E. Walter Maunder examined data on the carbon-14 content of tree rings (see Chapter 1) and concluded that sunspot activity almost completely died away between 1645 and 1715, the period now known as the Maunder minimum. During this time Europe experienced years of record low temperatures, often termed the Little Ice Age, and the western United States experienced a severe drought. During the Grand Maximum (1000–1250), when solar activity appeared to be greater than usual, the northern hemisphere appeared to be warmer than present. It was during this period that the Vikings colonized Greenland.

Geological evidence shows that, further back in time, the Earth experienced prolonged periods of major glaciation known as ice ages. These ice ages lasted a few million years and appear to have occurred every 250 million years since about 1.25 billion years ago. They may be caused when the Sun and Solar System pass through clouds of interstellar dust in the Galaxy, thereby dimming the Sun's light. A galactic year lasts approximately 250 million Earth years, adding weight to this theory.

Space-age problems

The top of the Earth's atmosphere reacts dramatically to fluctuations in solar output. Above the stratosphere, in the upper reaches of the mesosphere (above 50 to 100 km), the temperature varies between about 500 and 2000 K on a daily basis as the Earth turns on its axis and the atmosphere passes in and out of the Earth's shadow. As its temperature rises, the atmosphere expands, as any heated gas will do, and this can have an affect on any low-altitude spacecraft.

The orbits of spacecraft lying within the Earth's atmosphere will eventually decay because of the drag from the air. The more dense the air, the more the spacecraft will experience a drag and the quicker it will decay. This is something mission controllers have to plan for when sending expensive equipment into Earth orbit, and they will work out the atmospheric drag at the satellite's altitude.

If the temperature of the atmosphere increases, the atmosphere expands and the satellite experiences greater drag because it is traveling through denser air which has expanded up to meet it from below.

In addition to the daily variation in the temperature of the atmosphere, the Sun's activity level also has an effect: the more active the Sun at solar maximum, the more energy it emits and the more heat the upper atmosphere receives. This was demonstrated rather ironically by the Solar Maximum Mission satellite (see Chapter 9), which was launched in 1980 to observe the Sun during maximum activity. It was launched into an orbit of altitude 574 km, but because of the high solar activity the orbit quickly decayed to an altitude of about 500 km in 1983. The orbit decayed much more gradually over the next four years,

The Earth's magnetosphere

The solar wind, composed of charged particles, continually flows out from the Sun, carrying the solar magnetic field with it. This wind interacts with all bodies in the Solar System, causing cometary tails to stream out, away from the Sun, aurorae to form in planetary atmospheres, and planetary magnetic fields to distort. Flares and coronal mass ejections can cause terrific gusts in the solar wind, temporarily distorting the planetary magnetic fields further.

The Earth's magnetic field is a simple dipole, running very close to its axis of rotation. With no influences like the Sun's solar wind, the magnetic field lines would appear like those of a simple bar magnetic, but the solar wind pushes against the field lines as it impinges on the Earth. Thus the lines sunward are compressed, and the lines on the side of the Earth away from the Sun are stretched out.

At the point where the solar wind first meets the Earth's magnetic field, a bow shock is set up and the solar particles are redirected either side of the Earth. The region of space in which the Earth's magnetic field dominates the solar wind is known as its magnetosphere. The boundary of the magnetosphere is called the magnetopause, and the elongated field lines downstream from the Earth is called the magnetotail.

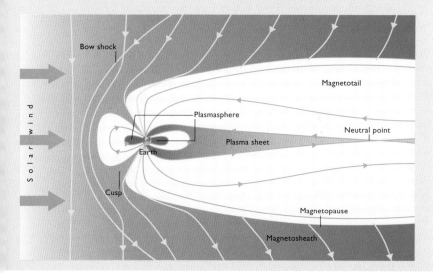

▲ As charged particles from the solar wind impinge on the Earth's magnetosphere, some particles spiral down the Earth's magnetic field lines at the poles. When they reach the atmosphere, they interact with atmospheric atoms and molecules causing the beautiful aurorae. In this image, interactions with oxygen give a red glow.

and after experiencing another increase in solar activity, the spacecraft burned up in the lower atmosphere in 1989.

Spacecraft and satellite designers have to take solar activity into account when planning space missions. In addition to the increased atmospheric drag that occurs during high solar activity, gusts in the solar wind can have a dramatic effect on both spacecraft and humans in space.

There have been many failures of spacecraft systems because of radiation or particle bombardment from the Sun. Fifty or so spacecraft or satellites have been damaged as a direct result of solar activity. One, the American telecommunications satellite Telstar 401, stopped transmitting as a direct result of a solar flare on January 11, 1997.

Earthly interaction with the solar wind

The Sun continually emits subatomic particles into space. This stream of charged particles is called the solar wind (see Chapter 1). When the particles reach Earth, they can have various effects. Most of these

The ozone hole

Most of the oxygen in the Earth's atmosphere exists in molecular form, consisting of two oxygen atoms bound together. This molecular oxygen is the form of oxygen that we breathe.

High in the atmosphere, especially in the stratosphere, the ultraviolet light from the Sun interacts with oxygen molecules, splitting them into their constituent oxygen atoms. A free oxygen atom is highly reactive, and before very long it will latch on to a surviving oxygen molecule to form ozone, which is a molecule containing three oxygen atoms. Ozone produced in this way is at its most abundant between altitudes of about 20 and 30 km, where it forms what is known as the ozone layer.

This ozone layer is very insubstantial: if it were all gathered together, it would make a layer only a few millimeters thick at sea-level. Even so, it is highly effective at blocking out rays from the longer-wavelength parts of the ultraviolet spectrum, which would be harmful to life were they to reach the surface. In the long term this UV radiation can cause genetic mutations, and in the short term it can cause skin cancers. A 10% decrease in stratospheric ozone over the UK would lead to about 20,000 extra cases of skin cancer every year.

During the 1980s and 1990s, the Earth's ozone layer lost several percent of its ozone, probably because of the release of industrial chemicals called chlorofluorocarbons (CFCs), which were formerly widely used in refrigerators and as propellants in aerosol sprays. The spread of these CFCs into the ozone layer encourages ozone to break down to molecular oxygen. The depletion of ozone is most marked at high latitudes, particularly in the Antarctic, where an annually varying ozone hole has been noted since the late 1970s.

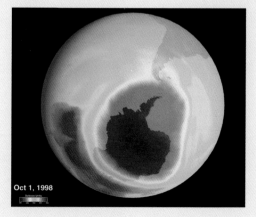

▶ In this image the blue/purple regions are areas of low ozone while the red regions show the highest percentage of ozone. The low ozone region is situated over Antarctica.

Oct 1, 1998

effects are benign, and unless you were specifically monitoring the Sun you would not be aware of them, but some are more noticeable.

The most visually obvious (and the most beautiful) are the aurorae. When solar particles penetrate Earth's atmosphere, they interact with atmospheric atoms and molecules to produce colored emissions which can dance in the upper layers of the atmosphere for days. These interactions happen between altitudes of about 100 and 1000 km, and are generally near the polar regions.

Aurorae are usually seen from terrestrial latitudes of between 60° and 75° in both hemispheres. This is because the charged solar particles can only gain access to the Earth's atmosphere by spiraling down the Earth's magnetic field lines near the two poles. The charged particles interact with the atmosphere in regions known as auroral ovals which surround the poles. The greater the strength of the solar wind, the lower (and larger) these ovals become.

Aurorae can be amazing sights: colorful glows that weave and flicker across the whole sky. Interactions with oxygen produce red and green photons, while interactions with nitrogen produce blue and violet lights.

At the top of the atmosphere

Energetic solar radiation penetrates the upper levels of the Earth's atmosphere. Luckily for life at the surface, most of the harmful radiation is blocked or absorbed before it reaches the ground (see the feature on page 129).

The solar radiation at X-ray and extreme ultraviolet wavelengths interacts with atoms and molecules in the upper atmosphere, ionizing them. This creates the ionosphere – layers of free electrons and ions produced by these interactions that lie between 60 and 1000 km above the Earth's surface.

The ionosphere is of great practical significance because it allows long-distance communication by high-frequency (short-wave) radio waves. Radio waves with frequency higher than about 20 MHz can be "bounced" off the ionosphere, thereby reaching around the curvature of the Earth, which would normally stop direct communication. During a solar storm or after the eruption of a solar flare, the ionosphere is enhanced and effectively moves upward. Radio signals then fail to reach the ionosphere, and are absorbed before they can undergo reflection. The resulting radio blackouts do not last long and are termed sudden ionospheric disturbances (SIDs).

Stormy space weather

A solar flare can release huge amounts of energy and a huge number of particles into the solar wind, causing strong gusts which envelop the

Earth and other members of the Solar System. Gusts of solar material called coronal mass ejections are also released (see Chapter 1).

In addition to solar storms affecting the Earth's ionosphere and causing disruption of short-wave radio communications, solar activity can also disrupt navigation systems on ships and aircraft, and military radar systems. During World War II, the British radar stations along the south coast were frequently knocked out of action by solar activity (the Sun was near maximum during the war).

Surges in long electricity transmission lines can cause widespread power blackouts, as happened in Quebec in March 1989, when 6 million people were left without electricity after a huge solar-induced magnetic storm. Damage to microchips and electrical discharges may also cause satellites to stop operating, causing disruption of, for example, telephone, television and data communication services.

Solar storms have many more unusual effects. With the temporary disruption of the Earth's magnetic field, birds, which fly using the magnetic field as a navigation aid, can get lost. In 1989, during a large pigeon race in America, over 70% of birds got lost because of the disruption caused by a solar flare. Solar storms have even been blamed for increased corrosion in oil pipelines as electrical currents are set up across them during magnetic storms.

9 · PROFESSIONAL SOLAR ASTRONOMY

The Sun has been observed professionally in white light for over two hundred years; the relative sunspot number has been recorded daily since 1848 (see Chapters 1 and 5). Since World War II, the Sun has been monitored in other wavelengths, such as radio, ultraviolet and hydrogen alpha. Now that solar instruments can be carried above the Earth's atmosphere on board spacecraft, the Sun can be observed across the whole electromagnetic spectrum, including in the higher-energy bands such as X-ray and extreme ultraviolet, which are very important for researching many solar phenomena.

The Sun is now constantly monitored from space, and daily images are available in a variety of wavelengths on the Internet. Daily data are also published, bringing the professional view of the Sun into the domain of the amateur astronomer and interested layman. One result of this ready availability is that images of the Sun from spacecraft such as SOHO have become more familiar to the general public. That familiarity has not, however, diminished the peculiar beauty of seeing our nearest star in eruption.

▼ The rather elegant McMath–Pierce Solar Telescope has a 152-meter shaft.

The optics of this telescope are aligned with the Earth's rotation axis.

▶ *The Big Bear Solar Observatory has been built near water, which helps to steady the Earth's atmospheric turbulence and hence aids in giving a steady solar image.*

Ground-based instruments

Professional telescopes for observing the Sun do not resemble ordinary telescopes as they are built to minimize the input of light, instead of maximizing it as observers of, for example, faint galaxies need to do. Solar telescopes are usually built at high altitudes because atmospheric turbulence decreases with height above the ground. The shorter the path of sunlight through the atmosphere, the better, so a telescope receiving light that has traveled vertically through the atmosphere creates a steadier image than one receiving sunlight that has traveled obliquely through the Earth's air.

The McMath–Pierce Solar Telescope at Kitt Peak, Arizona, operated by the National Solar Observatory, has its main optics aligned with the Earth's rotation axis. Its long focal length produces a solar image 0.85 meters across, and the image is projected by a rotating mirror (a heliostat) down a 152-meter shaft to a fixed mirror, 1.6 meters in diameter and 52 meters below ground. This in turn reflects the image back up to ground level, where the image is formed in an observation room containing spectrographs. As with other solar telescopes, it has its own architectural beauty.

Restricted viewing

Ground-based solar telescopes are restricted by the Earth's atmosphere in the wavelengths at which they can view the Sun. In general they are used to observe in white light or, coupled with a spectroheliograph, at a particular wavelength, or wavelength band, such as hydrogen alpha (see Chapter 1). Solar spectra are recorded, and some observatories have spectroheliographs which can record the movement of solar material toward and away from the Earth. These "Dopplergrams" show dark and light areas where solar material is either rising or falling, and gray areas which are at rest with respect to the Earth.

Magnetographs produce images (magnetograms) of the Sun's magnetic field, a very important aspect of all solar phenomena (see Chapter 1), and some observatories can study the corona by using coronagraphs. The study of the inner corona, however, is extremely difficult from Earth-based observatories because the shifting atmosphere can cause tiny portions of the photosphere to appear, thereby instantly drowning out the much fainter corona.

The Sun can also be observed at radio wavelengths from the ground. To improve resolution, interferometry is often used – a technique that uses several receivers coupled together to simulate a much larger receiving dish. Different arrays of radio dishes observe the Sun in different wavelengths. For example, the Very Large Array (VLA) in New Mexico observes the Sun at 2, 6 and 20 cm, while the Nançay Radioheliograph in France operates in the 10 cm to 1 meter range.

GONG

▲ The MDI instrument on board SOHO investigates solar rotation rates and the motion of the solar plasma. In this image, the yellow/orange/red areas are rotating faster than the blue areas.

Helioseismology is the study of internal solar seismic waves, using instruments that measure minute Doppler shifts in the wavelengths of absorption lines. Interpretation of the data they gather reveals information on temperature, chemical composition, and motions from just below the surface down to the very core of the Sun. Helioseismology can be done from the ground, but a major obstacle in observing from the ground is the interruption in the observations when night falls. This introduces uncertainties in the determination of the period of the waves, as well as creating background noise which hides all but the strongest oscillations. To overcome this problem, the Global Oscillation Network Group (GONG) has developed a network of instruments around the world to provide continuous observation.

The GONG data, along with helioseismology measurements made with the MDI instrument on board the SOHO spacecraft (see below), allow scientists to monitor the onset of active areas on the Sun and hence the possibility of space storms (see Chapter 8). It is even possible to determine what is happening on the Sun's far side by probing through the interior.

Observing the Sun from space

To observe the Sun in wavelengths other than visible light and some radio bands, instruments need to be above the Earth's atmosphere. This type of solar observation began with instruments carried on board rockets and high-altitude balloons. Today there are satellites specifically designed for solar observation, and windows into the Sun have been well and truly opened.

Past missions that had a significant solar observing program include Skylab, the Solar Maximum Mission, Spacelab 2 and Yohkoh. Missions that are still operating include Ulysses, SOHO and TRACE. (See the feature on pages 136.) Other missions are investigating the Sun–Earth connection: for example POLAR, which studies the Earth's aurorae and magnetosphere. Alongside missions specifically designed to study the Sun are satellites analyzing aspects of the solar wind and aspects of the Earth's magnetosphere. Together with solar observations, these spacecraft are gradually building up a picture of how the Sun affects the Earth and its environs.

Past missions

Skylab was a manned Earth-orbiting space station, launched on May 14, 1973 and used for general astronomical and solar observation. Skylab's instruments observed in X-rays, ultraviolet, white light and hydrogen alpha. Over 180,000 solar images were obtained, revealing for the first time how dynamic the Sun is, even over the course of a day, and improving our understanding of processes occurring in the outer solar atmosphere, and their links with photospheric and chromospheric phenomena.

The Solar Maximum Mission (SMM) satellite was launched on February 14, 1980, to observe the Sun at the most active part of the solar cycle. It operated for nearly 10 years, observing the Sun from the visible region to gamma-ray wavelengths. It found that the Sun is actually brighter at solar maximum, when there are more sunspots on the surface – although the spots are dark, they are surrounded by bright faculae. SMM studied the release of energy from flares, showing that gamma-ray emission from even modest flares is fairly common and that particles are accelerated in flares to high energies in just a few seconds.

Spacelab 2 was a package of instruments flown on board the space shuttle for a week, beginning July 29, 1985. The astronauts operated

Major solar space missions

PAST MISSIONS	
Mission	**Subject of study**
Skylab	The Sun in X-rays and the ultraviolet
Solar Maximum Mission (SMM)	Solar photosphere and corona
Yohkoh	Solar atmosphere at X-ray and gamma-ray wavelengths
Spacelab 2	Magnetic field in the photosphere
GOES (The Geostationary Operational Environmental Satellite)	The Sun in X-rays

ONGOING MISSIONS	
Mission	**Subject of study**
Polar	Multiwavelength imaging of Earth's aurorae and magnetosphere; measurements of solar wind content, strengths of magnetic and electric fields
TRACE (Transition Region and Coronal Explorer)	Solar corona
SOHO (Solar and Heliospheric Observatory)	Solar atmosphere at X-ray and gamma-ray wavelengths
Ulysses	Latitudes of the Sun above 70° N and S
Cluster	Solar wind and Earth's magnetic field
HESSI (High Energy Solar Spectroscopic Imager)	Images and spectra of the Sun in X-rays and gamma rays

PLANNED MISSIONS	
Mission	**Subject of study**
Solar Orbiter	Outer solar atmosphere and the heliosphere
Solar-B	Optical, X-ray and ultraviolet observations of the Sun
STEREO (Solar Terrestrial Relations Observatory)	Coronal mass ejections
Solar Probe	Solar corona and polar photosphere

◀ *Two astronauts in the cargo bay of the* Challenger *shuttle repair the Solar Maximum Mission Satellite (SMMS). The astronauts are George C. Nelson (right) and James D. van Hoften (left).*

four solar instruments. One of these instruments, the Solar Optical Universal Polarimeter, obtained high-resolution images of the solar granulation. The smallest features recorded measure 175 km across.

The satellite Yohkoh (Japanese for "sunbeam") was launched on August 31, 1991, to study flares and other solar activity in X-rays and gamma rays. Its instruments were designed jointly by Japanese, American and European scientists. One of Yohkoh's two X-ray telescopes imaged the Sun in the 0.4–6.0 nm wavelength range with a resolution down to 2″, the equivalent of 1700 km on the solar surface – far greater than had been possible with Skylab or SMM. The mission ended when, during the solar eclipse of December 14, 2001, the spacecraft lost its pointing capability and the batteries discharged.

Ulysses

The Ulysses probe, launched on October 6, 1990, was designed to fly over the poles of the Sun. To achieve this it first had to travel to Jupiter, where it received a gravitational "kick" which sent it out of the plane of the ecliptic. Ulysses first flew over the Sun's south pole during May–November 1994, and over the Sun's north pole during June–October 1995.

In 2004, Ulysses is still in operation. It has observed the Sun over a complete solar cycle, and has surveyed the high-latitude heliosphere within 5 astronomical units of the Sun over the full range of solar activity. Ulysses has confirmed many scientific predictions concerning the Sun's magnetic field at its poles, and the flow of solar material at these high latitudes, but has also delivered some surprising results.

It had been thought that the solar wind would increase in speed toward the poles, but Ulysses discovered that at high latitudes the solar wind speed remains fairly constant, at about 750 km/s.

Before Ulysses, the Sun's magnetic field was assumed to be similar to the Earth's dipole field, with lines near the solar equator forming closed loops, and lines near the poles dragged far into interplanetary space by the solar wind. For a dipole, the field strength over the poles is twice that at the equator, but Ulysses found that the Sun's magnetic field strength does not vary greatly with latitude. The data are still being analyzed.

SOHO

One of the most productive and best-known solar observatories currently in operation is the Solar and Heliospheric Observatory (SOHO). The observatory, a collaboration between ESA and NASA, was launched on December 2, 1995, to study the Sun's internal structure and its atmosphere, and to monitor the solar wind.

SOHO is able to monitor the Sun continuously because instead of orbiting the Earth, with its view blocked by the Earth for a period of time during each orbit, it orbits the inner Lagrangian point – a point between the Sun and the Earth where the gravitational attractions of the two bodies effectively cancel out. This point is 1.5 million km sunward from the Earth. At this distance, SOHO is also free from the effects of the Earth's atmosphere.

SOHO has returned a huge amount of data which is still being studied, data that has revealed a few more of the Sun's secrets. It has measured temperature, density, composition and velocities in the corona, and has followed the evolution of coronal structures at high resolution. By combining measurements of velocity oscillations of the full solar disk with measurements of the solar energy output across the face of the Sun, SOHO can investigate the solar nucleus. High-resolution measurements of oscillations in the photosphere give precise information about the Sun's convective zone – the outer layer of the solar interior.

As well as monitoring the solar wind, SOHO maps the hydrogen density in the heliosphere. By using telescopes sensitive to a particular wavelength of hydrogen, the large-scale structure of the solar wind streams can be measured. The data from these instruments are used in conjunction with data from Cluster (see below) and Ulysses.

SOHO was designed to operate until 1998, but it has been so successful that ESA and NASA decided to prolong its life. This extension enabled project scientists to compare the Sun's behavior from when it had few sunspots (1996) to the peak of sunspot activity around 2000. It has also allowed SOHO to work, as planned, alongside the delayed Cluster mission.

Signals from SOHO disappeared for several weeks in 1998, but communication was recovered and it was brought back into operation after a gap of more than four months. Other difficulties came with the loss of the gyroscopes used to control the spacecraft's orientation. Despite these problems, engineers have kept SOHO functioning with all its instruments performing well.

Cluster

Cluster is an ESA mission consisting of four satellites flying in a tetrahedral formation, giving a three-dimensional view of the solar wind and of the interaction between the Sun and Earth (see Chapter 8). The original Cluster mission was lost during launch on June 4, 1996, when the Ariane-5 rocket, being used for the first time, exploded. ESA agreed to fund a replacement mission, which was successfully launched in the summer of 2000 on two Russian Soyuz rockets from Baikonur Cosmodrome in Kazakhstan.

Cluster II's goals and instruments are the same as those of the original mission, and it works together with SOHO and Ulysses. SOHO constantly watches the Sun, monitoring its activity; Cluster monitors the effects of the solar wind gusts which buffet the magnetic field of the Earth; and Ulysses patrols the Sun in a tilted orbit, well away from the plane of the planets, obtaining a more global view of the solar wind. The Cluster satellites can move close together with only 100 km between them to get a concentrated view of the solar wind, and can pull away from each other as far as 20,000 km to get the bigger picture.

The mission has confirmed that the outer regions of the Earth's magnetosphere are constantly being rocked by big "waves" in the solar wind. Cluster has also shed light on so-called black aurora: the strange electrical phenomenon that produces dark, empty regions within the visible aurorae seen at both poles (see Chapter 8). The black aurora takes on various forms: dark rings, curls or black blobs in a sea of faint, glowing aurora. Cluster has shown that these peculiar patches occur where there are holes in the ionosphere, the part of the upper atmosphere where aurorae are created. Here, the particles that make up the ionosphere are shooting upward into space inside regions termed "positively charged electric potential structures." This is the opposite of the process that creates visible aurorae, where electrons spiral down from space into the atmosphere within similar, but negatively charged, structures.

TRACE

The Transition Region and Coronal Explorer (TRACE) was launched on April 2, 1998, on a Pegasus launch vehicle from Vandenberg Air Force Base to make joint observations with SOHO. TRACE studies

▶ This image, captured by the TRACE satellite, shows solar material constrained by the solar magnetic field into impressive coronal loops arching high above the solar photosphere.

the magnetic field in the solar atmosphere, observing in the ultraviolet and extreme ultraviolet parts of the electromagnetic spectrum.

One of the mysteries of the corona, the outer atmosphere, is that its temperature is more than a hundred times higher than the photosphere below it. It is now known that it is the Sun's magnetic fields that gives the corona its characteristic shape, dominated by arches or loops of different brightness and temperature, but the process by which it is heated has long eluded solar physicists. It is obvious that a lot of energy goes into heating these layers, but it is unclear how it gets there, or even where it is predominantly deposited.

Recent analysis of data from TRACE and SOHO reveals that most of the heating occurs in the first 10,000 to 20,000 km above the surface. This is only a fraction of the altitude of some of the bright coronal structures, many of which arch several hundred thousand kilometers above the Sun. Moreover, not only is heat deposited primarily low down, but also the coronal gas in many places is actually thrust upward at velocities that can reach 300,000 km/h.

Future missions

The Japanese Solar-B mission is a follow-up to the highly successful Yohkoh. The mission will carry a coordinated set of optical, ultraviolet and X-ray instruments to investigate the interaction between the Sun's atmosphere and its magnetic field. It should improve our understanding of what creates the driving force behind space weather.

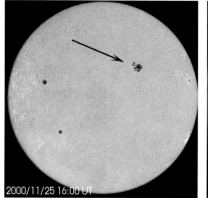

▲ In these two images taken by instruments on board SOHO, the sunspot group seen in white light on the left is shown as a bright white spot in X-ray in the right-hand image. This sunspot group was the source of a vast coronal mass ejection, the progress of which is shown in images on the next page.

NASA's Solar Terrestrial Relations Observatory (STEREO) is planned to investigate the enigmatic coronal mass ejections. A US–Russian collaboration is the Solar Probe, which will be sent deep into the solar corona – far closer to the Sun than any other spacecraft has previously ventured. Another planned mission is the ESA's Solar Orbiter, which will be sent out of the plane of the ecliptic to make observations of the inner heliosphere and the poles of the Sun.

Pro-am collaboration

Many amateur astronomers observe the Sun regularly. Some attempt to keep a daily record (weather permitting), and some observe with very sophisticated equipment allowing them to gauge the level of solar activity by observing phenomena such as flares, filaments and prominences. There has long been a close connection between the amateur and the professional solar astronomer.

Before the onset of solar space missions, the only means of observing the Sun continuously was from the ground – again weather permitting. Since amateurs could count the number of sunspots with fairly modest equipment, and they observe from all over the globe, professionals made use (and still make use) of relative sunspot number counts from amateurs. To contribute to this database, you can send your counts directly to amateur organizations such as the British Astronomical Association or the American Association of Variable Stars Observers. Alternatively, you can join a local society that contributes to the program.

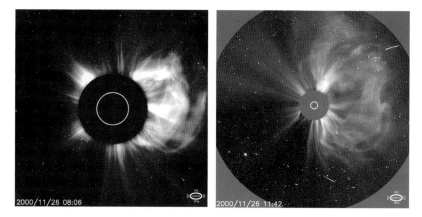

▲ In these two images taken by the LASCO instrument on board the SOHO spacecraft, the flare emitted by the sunspot group shown in white light on the previous page has developed into a massive coronal mass ejection. It can be seen billowing out thousands of miles from the Sun.

In more recent times, more active programs of pro-am collaboration have been instituted at a few observatories. For example, some amateurs use the coronagraphs at the Pic du Midi Observatory in France. Today, amateurs have daily access to the data produced by space observatories such as SOHO. They can use the data to enhance their own observing programs or to undertake their own research programs.

As professional research has become more readily accessible to the amateur, it has changed the way some amateurs approach solar observing. Today many amateurs have an up-to-date knowledge of the latest solar research, which enhances their understanding of what they observe and record. They can view real-time professional images of the Sun and can compare them with their own observations, allowing them to interpret their results far more fully than ever before, especially as they have access to solar images taken in other wavelengths, showing different solar phenomena.

The Internet has widened the scope of solar observing, not only by giving the amateurs access to professional information and images, but also allowing everyone the opportunity to see the Sun every day, even if it is cloudy where they live.

— FORTHCOMING SOLAR ECLIPSES —

This table lists solar eclipses up to the year 2009. The start and finish times are appropriate to the location quoted, but this time is not necessarily the start and end time of the eclipse if some portion occurs beneath the horizon. Only major locations are given. Even if an eclipse is total, it will not be total from every location mentioned. All times are quoted in Universal Time.

OCTOBER 14, 2004: PARTIAL

A partial eclipse crossing northeastern Asia, reaching more than 90% over Bethel, Alaska.

Location	Type of eclipse	Times: Start	Mid	End
Japan (Tokyo)	Partial (24.0%)	01:45	02:40	03:36
Alaska (Bethel)	Partial (92.4%)	01:50	02:57	sunset

APRIL 8, 2005: TOTAL/ANNULAR

This is a very rare type of eclipse with the Moon being almost exactly the same size as the Sun in the sky. A total eclipse is produced at the center of the eclipse path, but at the extreme ends, just as the Moon's shadow reaches and leaves the Earth's surface, an annular eclipse occurs. The total/annular track is narrow, but a partial eclipse will be seen from a much larger area of the Earth's surface.

Location	Type of eclipse	Times: Start	Mid	End
Panama (San Carlos)	Total/Annular	20:56	22:12	23:18
Panama (Panama City)	Partial (98.5%)	20:57	22:12	23:18
Colombia (Pueblonuevo)	Total/Annular	21:02	22:15	23:18
USA (Houston)	Partial (29.5%)	21:17	22:11	23:01
USA (Miami)	Partial (46.6%)	21:20	22:19	23:14
USA (New Orleans)	Partial (31.4%)	21:21	22:15	23:05
New Zealand (Auckland)	Partial (65.6%)	sunrise	05:49	06:49

OCTOBER 3, 2005: ANNULAR

This will be a broad annular eclipse seen across a large part of Africa, with the Moon being very much smaller than the Sun as it appears in the sky. A partial eclipse will be visible from many European cities.

Location	Type of eclipse	Times: Start	Mid	End
England (London)	Partial (66.3%)	06:48	08:01	09:18
Portugal (Braga)	Annular	07:38	08:53	10:16
Spain (Madrid)	Annular	07:40	08:57	10:23
Algeria (Sétif)	Annular	07:46	09:10	10:41
France (Paris)	Partial (70.2%)	07:47	09:02	10:22
Italy (Rome)	Partial (73.4%)	07:53	09:15	10:43
Germany (Munich)	Partial (61.3%)	07:56	09:11	10:32
Greece (Athens)	Partial (63.7%)	08:09	09:34-	11:03

Kenya (Marsabit)	Annular	09:31	11:16-	12:51
Somalia (Chisimayu)	Annular	09:48	11:30-	13:01
South Africa (Johannesburg)	Partial (14.6%)	11:01	11:50-	12:37

MARCH 29, 2006: TOTAL

This total eclipse will be visible across Africa with a partial eclipse visible from some major cities of Europe. If you visit Natal in Brazil, you will see the Sun rise in eclipse.

Location	Type of eclipse	Times: Start	Mid	End
Brazil (Natal)	Total	sunrise	08:35	09:34
Ghana (Accra)	Total	08:00	09:11	10:29
Nigeria (Gusau)	Total	08:20	09:33	10:54
Niger (Maradi)	Total	08:20	09:36	10:58
Spain (Madrid)	Partial (35.7%)	09:17	10:12-	11:09
Italy (Rome)	Partial (59.5%)	09:27	10:36-	11:45
Egypt (Cairo)	Partial (86.4%)	09:27	10:47-	12:06
France (Paris)	Partial (33.8%)	09:39	10:32-	11:26
England (London)	Partial (27.5%)	09:45	10:33	11:21
Turkey (Sivas)	Total	09:50	11:07	12:21
Russia (Moscow)	Partial (65.1%)	10:10	11:15	12:18

SEPTEMBER 22, 2006: ANNULAR

A large part of the track of this annular eclipse lies across the Atlantic, but starts from north-eastern South America.

Location	Type of eclipse	Times: Start	Mid	End
French Guiana (Cayenne)	Annular	sunrise	09:52	11:10
Brazil (Oiapoque)	Annular	sunrise	09:53	11:12
Brazil (Rio de Janeiro)	Partial (39.5%)	09:41	10:47-	12:02
South Africa (Cape Town)	Partial (70.1%)	11:20	12:56	14:24
South Africa (Johannesburg)	Partial (38.1%)	11:43	13:03-	14:15

MARCH 19, 2007: PARTIAL

This partial eclipse will cover central and eastern Asia and most of western Alaska.

Location	Type of eclipse	Times: Start	Mid	End
China (Beijing)	Partial (39.5%)	01:27	02:23	03:21
India (Delhi)	Partial (39.5%)	sunrise	01:37	02:31
Alaska (Nome)	Partial (10.3%)	03:25	03:51-	04:17
Korea (Seoul)	Partial (19.5%)	01:47	02:31	03:15

SEPTEMBER 11, 2007: PARTIAL

This partial eclipse will be visible from some areas of South America.

Location	Type of eclipse	Times Start	Mid	End
Paraguay (Asunción)	Partial (38.9%)	10:31	11:30	12:35
Brazil (Rio de Janeiro)	Partial (25.7%)	10:43	11:39	12:39

Chile (Santiago)	Partial (55.0%)	sunrise	11:39	12:48
Colombia (Buenos Aires)	Partial (51.9%)	10:41	11:48	13:01
Falkland Islands (Stanley)	Partial (65.3%)	11:13	12:25	13:42
Chile (Cape Horn)	Partial (70.9%)	11:18	12:27	13:42

AUGUST 1, 2008: TOTAL

This total eclipse crosses some inhospitable parts of the Earth's surface, starting in the far north of Canada, crossing the Arctic Ocean and the coast of Greenland before crossing to northern Russia. It ends in China. A partial eclipse will be visible from some major European cities.

Location	Type of eclipse	Times: Start	Mid	End
England (London)	Partial (21.8%)	08:32	09:17	10:04
Northwest Territories (Alert)	Total	08:36	09:32	10:29
France (Paris)	Partial (14.3%)	08:42	09:20	10:00
Germany (Berlin)	Partial (30.0%)	08:43	09:37	10:32
Russia (Moscow)	Partial (57.9%)	09:01	10:09	11:14
Russia (Nadym)	Total	09:16	10:21	11:23
Russia (Novosibirsk)	Total	09:41	10:45	11:45
Mongolia (Bulgan)	Total	10:01	11:02	11:59
China (Lanzhou)	Partial (99.2%)	10:23	11:19	–
India (Delhi)	Partial (62.8%)	10:32	11:31	12:26

JANUARY 26, 2009: ANNULAR

A large part of the track of this annular eclipse lies across water.

Location	Type of eclipse	Times: Start	Mid	End
South Africa (Cape Town)	Partial (72.0%)	04:58	06:11	07:37
Australia (Perth)	Partial (33.6%)	08:01	09:01	09:55
Java (Serang)	Annular	08:19	09:40	10:50
Sumatra (Teluk Betung)	Annular	08:19	09:41	10:51
Singapore (Singapore)	Partial (79.6%)	08:29	09:49	10:57
Borneo (Sampit)	Annular	08:31	09:46	10:52

JULY 1, 2009: TOTAL

Asia is again well placed for this total eclipse, the path of which crosses India and China. The city of Wuhan lies directly under the path of totality and will experience a totality of more than 5 minutes.

Location	Type of eclipse	Times: Start	Mid	End
India (Jabalpur)	Total	23:59	00:53	01:54
India (Patna)	Total	23:59	00:56	01:59
Bhutan (Thimpu)	Total	00:00	00:59	02:04
China (Chengdu)	Total	00:06	01:12	02:26
China (Wuhan)	Total	00:14	01:26	02:46
Japan (Tokyo)	Partial (74.8%)	00:55	02:12	03:30
Hawaii (Honolulu)	Partial (11.6%)	03:21	03:47	04:13

GLOSSARY

absorption line A dark line in a continuous spectrum caused by photons of a specific wavelength being absorbed.

altazimuth A telescope mount that has one axis parallel to the horizon (azimuth axis) and the other axis at right angles to the horizon (altitude axis).

Ångström (Å) A unit of length equal to 10^{-10} m, often used to express the wavelength of light.

annular eclipse An eclipse of the Sun in which the Moon's apparent diameter is not large enough to cover the solar photosphere completely, thereby leaving a ring or "annulus" of photospheric light around the lunar disk.

aperture The diameter of the observing lens or mirror of a telescope.

aphelion The point at which a body in orbit about the Sun lies at its farthest from the Sun.

arcminute (') A unit of angular measure; 60 arcminutes = 1 degree.

arcsecond (") A unit of angular measure; 60 arcseconds = 1 arcminute.

astronomical unit (symbol AU) The mean distance of the Earth from the Sun. 1 AU equals 149,597,870 km.

atom The smallest independent constituent of an element, consisting of one or more protons and (except for hydrogen) neutrons in a nucleus with a surrounding cloud of electrons.

aurora A colored glow in the Earth's atmosphere, generally seen from high latitudes, caused by solar wind particles entering the upper atmosphere along the Earth's magnetic field lines and interacting with gases in the atmosphere.

Baily's beads An effect observed during a total solar eclipse when the Moon's limb lies close to the Sun's limb, allowing photospheric light to shine through lunar valleys.

Balmer series A series of absorption or emission lines in the spectrum of hydrogen produced by electrons transferring to and from the second energy level in the hydrogen atom.

bandwidth The range of frequencies over which an instrument is sensitive.

bipolar pair Name given to a pair of sunspots on account of their different magnetic polarities.

bright point A small bright region in the umbra of a sunspot. Bright points can form a network across the umbra similar in structure to the photospheric granulation.

Carrington rotation The mean sidereal rotation period of the Sun defined by Richard Carrington as 25.38 days, used as a basis for keeping track of successive rotations of the Sun.

central meridian The line of longitude that crosses the center of the solar disk from the north pole to the south pole.

chromosphere The layer of the Sun above the photosphere and below the corona.

CME Abbreviation for *coronal mass ejection*.

convective zone The region of the Sun in which convection is the main means of heat transfer.

core The innermost part of the Sun, in which hydrogen is being changed into helium by nuclear fusion.

corona The outer atmosphere of the Sun, visible during the totality of a solar eclipse.

coronal hole A dark region in the solar corona through which solar material is ejected into interplanetary space.

coronal mass ejection (CME) A massive ejection of solar material into interplanetary space.

diamond ring An effect observed during a total solar eclipse when the Moon's limb is very close to the Sun's limb, allowing photospheric light to shine through only one lunar valley.

differential rotation Rotation of a non-solid body in which different latitudes of the body rotate at different rates.

diffraction grating A device used to disperse light into a spectrum by means of a series of fine, closely spaced grooves ruled on to a flat surface.

dipole A pair of opposite magnetic poles.

eclipse The passage of a body through the shadow of another.

ecliptic The plane of the Earth's orbit around the Sun and the apparent path of the Sun against the background stars due to the Earth's orbital motion.

electromagnetic radiation Radiation emitted as an electromagnetic wave, generated by electric and magnetic fields oscillating at right angles to each other. Light is a form of electromagnetic radiation.

electromagnetic spectrum The complete range of *electromagnetic radiation*.

electron A negatively charged elementary particle. Electrons surround the nucleus of an atom.

emission line A bright line in a continuous spectrum caused by photons of a specific wavelength being emitted.

equinox Either of two points where the Sun's path crosses the celestial equator, or the dates on which this Sun is at one of those points (around March 21 and September 23).

eyepiece A system of lenses in a telescope through which an observer views the image.

faculae Bright hot clouds of solar material lying above sunspots in the upper photosphere.

filament A huge outpouring of solar material seen against the disk of the Sun; the same phenomenon viewed at the limb is a *prominence.*

filter A device that screens certain parts of the *electromagnetic spectrum.*

finder Abbreviation of finderscope, which is a small telescope mounted on a larger telescope and used to help point the larger telescope at the correct object.

first contact The point during a solar eclipse when the limb of the Moon first touches the solar disk.

first quarter The phase of the Moon, approximately seven days after new Moon, when half of the lunar disk is illuminated, and half is in darkness.

flare An enormous release of solar energy, usually associated with a sunspot.

focal length The distance between the center of the main observing lens or mirror in a telescope and its focal point (focus).

following spot (f-spot) The main spot in a sunspot pair or group which appears to follow behind in the direction of solar rotation.

fourth contact The point during a solar eclipse when the limb of the Moon leaves the solar disk.

Fraunhofer line An *absorption line* in the solar spectrum.

full Moon The phase of the Moon occurring when it lies opposite the Sun in the sky and hence appears fully illuminated.

fusion, nuclear The process by which hydrogen is transformed into helium at the center of the Sun.

gamma rays The most energetic type of electromagnetic radiation, with the shortest wavelength (less than 0.01 nm).

gauss (symbol G) A unit of magnetic field strength.

granulation The tops of solar convection cells seen in the photosphere as a network of lozenge shapes.

graticule A grid system used to observe or record the positions of solar features on the solar disk.

heavy element Any element heavier than hydrogen or helium.

heliographic latitude The system of latitude on the solar disk; the angular distance north or south of the Sun's equator.

heliographic longitude Longitude as measured on the solar disk: the angular distance from a standard or "zero" meridian, now defined as the meridian that passed through the ascending node of the Sun's equator on the ecliptic at 12:00 UT on January 1, 1854.

helioseismology The study of the structure of the solar interior by observing the sound waves that propagate through the photosphere.

helium (symbol He) The second lightest and second most abundant element in the Sun (and in the universe); an atom of helium consists of two protons and two neutrons in the nucleus, with two electrons.

hydrogen (symbol H) The lightest and most abundant element in the Sun (and in the universe); an atom of hydrogen consists of one proton in the nucleus, with a single electron.

hydrogen alpha (H-alpha, hydrogen-α) The spectral line at 656.3 nm, produced by electrons moving between energy levels 2 and 3 in the hydrogen atom.

hydrogen burning The fusion of hydrogen into helium in the cores of stars.

infrared Electromagnetic radiation with wavelength of between about 0.8 micrometers and 1000 micrometers.

inner bright ring A bright, narrow zone between the umbra and penumbra of a sunspot.

interferometry The observation of astronomical objects using interference patterns to give greater resolution.

ion An atom which has more or fewer electrons than in its normal, electrically neutral state.

ionization The process whereby a neutral atom gains or loses electrons, thereby gaining a negative or positive charge.

irradiance Another term for the *solar constant.*

joule (symbol J) A unit of energy.

kelvin (symbol K) A unit of temperature, or temperature difference. The zero point on the kelvin scale is equivalent to a temperature of $-273.15°C$.

Lagrangian point A point where the gravitational attractions between one body in orbit around another cancel out, thereby allowing a smaller body at that point to remain in equilibrium.

last quarter The phase of the Moon, approximately 21 days after new Moon, when half of the lunar disk is illuminated and half is in darkness.

light bridge A region spanning both the umbra and penumbra of a sunspot, effectively splitting a sunspot into two.

limb The edge of the visible disk of a spherical body such as the Sun.

limb darkening The effect observed at the limb of the Sun, a result of the observer looking obliquely through the top layers of the solar photosphere, whereas when the observer looks at the center of the solar disk they are seeing through the photosphere to the hotter, brighter solar interior.

magnetic solar cycle A period of about 22 years, twice the length of the solar cycle; the magnetic polarity in the solar hemispheres reverses from one solar cycle to the next.

magnetogram A diagram that shows the distribution strength and polarity of magnetic fields across the Sun's disk. It is taken by a magnetograph.

magnetosphere The region surrounding a planet such as the Earth with a magnetic field, in which the planet's magnetic field dominates that of the Sun.

magnitude (of a solar eclipse) The percentage of the Sun's disk covered by the Moon.

mean daily frequency The daily value of the relative sunspot number averaged over (generally) the period of a month.

molecule An atom or a number of atoms that are capable of independent existence.

nanometer (symbol nm) A unit of length equivalent to one-thousand-millionth of a meter (10^{-9} m), often used to express the wavelength of light.

neutron A neutrally charged elementary particle of similar mass to a proton, existing in the nuclei of atoms.

new Moon The phase of the Moon occurring when it lies between the Sun and Earth; only its far side is illuminated by the Sun and hence it is not visible from Earth at this time.

node Either of the two points at which an orbit intersects some reference plane. For example, the nodes of the Moon's orbit are where it crosses the plane of the ecliptic. The line of nodes is the line connecting these points.

nucleus The central, most massive part of an atom, containing protons and neutrons, around which electrons orbit.

observing blank A predrawn form for recording an observation of an object with a visible disk. A blank for solar observing consists of a predrawn circle with the diameter of the projected image of the Sun's disk, plus spaces for entering information such as observer, date, time, conditions and solar parameters.

optical density A measure of how much or how little visible light can pass through.

outer bright ring A region on the outside of the penumbra of a sunspot, separating the penumbra from the surrounding photosphere.

partial eclipse An eclipse in which one body is only partly covered by the shadow of another; a partial solar eclipse occurs when only part of the solar disk is observed to be covered by the Moon.

penumbra The lighter, grayish surroundings of the umbrae of larger sunspots. In a solar eclipse, the less dark outer region of the shadow, from where a partial eclipse can be seen.

perihelion The point at which a body in orbit about the Sun lies at its nearest to the Sun.

photon An individual "packet" of electromagnetic radiation with a specific energy.

photosphere The outer layer of the Sun that emits radiation in the visible part of the electromagnetic spectrum; sometimes termed the "surface" of the Sun.

pinhole camera A simple device using a small pinhole to project an image of the solar disk for observation.

plage A bright, hot cloud of chromospheric material associated with a sunspot, observed in hydrogen-alpha light.

plasma A state of matter composed of ionized atoms.

plume A short streamer emitted from a polar region of the Sun's corona.

polar faculae Hot, bright clouds of hydrogen, often circular in appearance, observed above the photosphere at high latitudes near the poles of the Sun.

pore A small, single dark area on the Sun less than 10 arcseconds across and with no penumbra (compare sunspot).

position angle In solar observing, the angle at which the north point of the rotation axis of the Sun is tilted west or east of north; it is negative if tilted toward the Sun's western limb, and positive if tilted toward the eastern limb.

preceding spot (p-spot) The main spot in a sunspot pair or group which appears to lead in the direction of solar rotation.

prominence A huge outpouring of solar material seen at the limb of the Sun.

proton A positively charged elementary particle of similar mass to a neutron, existing in the nuclei of atoms.

radiative zone The region in the Sun where radiation is the dominant form of heat transfer.

reflecting telescope (reflector) A telescope that uses mirrors to collect and focus the light from objects.

refracting telescope (refractor) A telescope that uses lenses to collect and focus the light from objects.

relative sunspot number (symbol R) An empirical formula used to indicate the degree of solar activity, based on counts of the number of sunspots and groups visible on the Sun's disk.

saros The period of revolution of the line of nodes (6585.32 days), after which a similar sequence of solar and lunar eclipses recurs.

second contact The point during a total solar eclipse when the leading (eastern) edge of the Moon touches the eastern limb of the solar disk and totality begins.

seeing The degree to which a telescopic image is distorted by turbulence in the Earth's atmosphere.

shadow bands Mottled, wavering patterns of dark and light which briefly pass across the ground during a total solar eclipse.

sidereal With respect to the stars.

SLR (camera) Abbreviation for single-lens reflex (camera), one in which the main lens is used both to capture and to view the image.

solar constant (irradiance) The total amount of solar radiation per unit area per unit time received at the top of the Earth's atmosphere (measured in watts per meter squared).

solar cycle The time period (about 11 years) over which solar activity varies between maximum and minimum, and back to maximum again.

solar disk The circular image of the Sun.

solar maximum The point in the solar cycle when the activity of the Sun is at a maximum.

solar minimum The point in the solar cycle when the activity of the Sun is at a minimum.

solar wind The constant flow of atomic particles from the Sun.

solstice Either of two points where the Sun is at its greatest distance north or south of the celestial equator, or the dates on which this Sun is at one of those points (around June 21 and December 22).

spectroheliogram A record of the solar spectrum.

spectroheliograph An instrument that images the Sun in a particular wavelength.

spectrohelioscope An instrument that allows the Sun to be observed directly in a particular wavelength.

spectroscope An instrument that produces a spectrum.

spectrum The distribution of radiation from an object such as the Sun.

spicule A column of solar material rising from the photosphere observed in hydrogen-alpha radiation.

Stoneyhurst disk One of a set of disks showing longitude and latitude of the Sun as viewed at different times during Earth's orbit.

stopping down The technique of creating a smaller aperture for a telescope, thereby reducing the amount of light (and heat) from the Sun passing through the instrument.

streamer An elongated structure in the Sun's corona.

sunspot In general, any dark area on the Sun; more correctly, a dark umbral region greater than 10 arcseconds in extent, or any region showing an umbra and penumbra (compare *pore*).

synodic With respect to two celestial bodies (for example the Earth and Sun).

third contact The point during a total solar eclipse when the trailing (western) edge of the Moon leaves the western limb of the solar disk and totality ends.

total eclipse An eclipse in which one body is totally covered by the shadow of another; a total solar eclipse occurs when the lunar disk completely covers the Sun's photosphere.

umbra The dark, inner region of a sunspot. In a solar eclipse, the dark inner region of the shadow, from where a total solar eclipse can be seen.

umbral dot A tiny dot in an umbra slightly brighter than the surrounding umbra.

universal time (UT) A timescale based on the rotation of the Earth. Since the Earth's rotation is not constant, this timescale is also not constant. It is equivalent to Greenwich Mean Time and is used for astronomical recording.

visible light Another term for white light.

void area An area in which the solar granulation is missing and which appears slightly darker than the surrounding photosphere; often a precursor to pores.

watt (symbol W) A unit of power.

wavelength The distance between two corresponding points on a wave (e.g. from one peak to the next).

white light (visible light) Electromagnetic radiation across all wavelengths observable by the human eye.

X-rays Electromagnetic radiation with wavelengths of between about 0.012 nm and 12 nm.

RESOURCES

Books

Beck, Rainer, Heinz Hilbrecht, Klaus Reinsch, and Peter Völker, eds. *Solar Astronomy Handbook*. Richmond, Virginia: Willmann-Bell, 1995.

Brody, Judit. *The Enigma of Sunspots*. Edinburgh, Scotland: Floris Books, 2002.

Dickinson, Terence. *NightWatch*. 3rd ed. Buffalo, New York: Firefly Books, 1998.

Dickinson, Terence, and Alan Dyer. *The Backyard Astronomer's Guide*. Rev. ed. Buffalo, New York: Firefly Books, 2002.

Guillermier, Pierre, and Serge Koutchmy. *Total Eclipses*. New York: Springer-Verlag, 1999.

Spence, Pam, ed. *The Universe Revealed*. New York: Cambridge University Press, 1998.

Zirin, Harold. *Astrophysics of the Sun*. New York: Cambridge University Press, 1988.

Annual publication
The Astronomical Almanac, Defense Department, U.S. Naval Observatory, Nautical Almanac Office.

Websites

American Association of Amateur Astronomers
www.corvus.com

American Association of Variable Star Observers (AAVSO)
www.aavso.org

American Astronomical Society
www.aas.org

Big Bear Solar Observatory
www.bbso.njit.edu

British Astronomical Association (BAA)
www.britastro.org

Cluster
spdext.estec.esa.nl/home/cluster

Global Oscillation Network Group (GONG)
gong.nso.edu

National Solar Observatory
www.nso.edu

Eclipse pages, NASA/Goddard Space Flight Center
sunearth.gsfc.nasa.gov/eclipse/eclipse.html

Royal Astronomical Society of Canada
www.rasc.ca

Skylab
science.nasa.gov/ssl/pad/solar/skylab.htm

SOHO
sohowww.nascom.nasa.gov

Solar-B mission
science.nasa.gov/ssl/pad/solar/solar-b.stm

Solar Influences Data Center (SIDC)
sidc.oma.be/index.php3

Solar Orbiter
www.orbiter.rl.ac.uk

Solar Probe
umbra.nascom.nasa.gov/solar_connections/probe.html

Solar-Terrestrial Relations Observatory (STEREO)
stereo.jhuapl.edu

Space Environment Center
www.sec.noaa.gov

Ulysses
ulysses.jpl.nasa.gov

Yohkoh
science.nasa.gov/ssl/pad/solar/yohkoh.htm

Software

For *Helio* (Peter Meadows' programs for working out heliographic latitude and longitude and area of sunspots) and other information for solar observers, visit www.meadows3.demon.co.uk.

ACKNOWLEDGMENTS

Abbreviations

AURA – Association of Universities for Research in Astronomy, Inc.

FTS – Fourier Transform Spectrometer (Kitt Peak)

SOHO – Solar and Heliospheric Observatory. SOHO is a project of international cooperation between the European Space Agency and NASA

NASA – National Aeronautics and Space Administration

NOAO – National Optical Astronomy Observatory

NSF – National Science Foundation

NSO – National Solar Observatory

Big Bear Solar Observatory 133

Bruce Hardie 46

Coronado Instruments 33

Courtesy of SOHO/EIT consortium 14t, 14b, 70r, 134

Courtesy of SOHO/MDI consortium 16

Courtesy of SOHO/UVCSEIT consortium 12b

Courtesy of SOHO/MDI/EIT/LASCO consortium 140–141

G. Scharmer/Swedish Vacuum Solar Telescope 9

H.J.P. Arnold/Sol Invictus 17b, 21, 22, 32, 34, 94t, 94b, 95, 96, 97, 98t, 98b, 99, 100, 101, 116, 117, 128

John Chapman-Smith/Galaxy 41b

NASA 129, 136

Nigel Sharp/NOAO/NSO/Kitt Peak FTS/AURA/NSF 31

NOAO/AURA/NSF 132

Pam Spence 37, 38, 39, 43, 44, 49, 65, 68, 69, 70l, 72, 76, 81, 83, 86, 110l, 110r, 112

Robin Scagell/Galaxy 41t

SEC/NOAA/US Department of Commerce 35

Till Credner, AlltheSky.com 12t

TRACE Project/NASA 139

Artwork by Raymond Turvey © Philip's

7940